Acknowledgements

The preparation of this report would not have been possible without the co-operation of the sand and gravel industry, particularly the marine sector; maritime local authorities; port authorities; Alluvial Mining & Shaft Sinking Co Ltd; Mr A D Bates (dredging consultant); Blue Circle Industries PLC; British Geological Survey; British Rail; British Standards Institution; British Telecom, Marine Division; Building Research Establishment; Cement & Concrete Association; Crown Estate Commissioners; Department of Energy, Petroleum Engineering Division; Department of Trade & Industry; Department of Transport; Foreign & Commonwealth Office; Hydaulics Research Ltd; Ministry of Agriculture, Fisheries & Food; National Maritime Museum; Messrs Sandberg; and D K Symes Associates.

The maps in Appendix 3 are based upon the Ordnance Survey's 1:10,000 scale maps with the permission of the Controller of Her Majesty's Stationery Office. Crown copyright reserved. Licence No. AL 813397.

The opinions expressed in this report are those of the Authors and do not necessarily coincide with those of the Department of the Environment.

Minerals Planning Research Project No. PECD 7/1/163-99/84

© Copyright Department of the Environment, 1986.

Front cover: Civil and Marine's dredger m.v. Cambourne discharging sand and gravel at the Company's Purfleet bulk materials depot. (Skyfotos Ltd.)

CONTENTS

DEPARTMENT OF THE ENVIRONMENT MINERALS DIVISION

DREDGING FOR SAND AND GRAVEL

Minerals Planning Research Project No. PECD 7/1/163–99/84

Patrick C.H. Chillingworth
 BA (Hons) FIQ MIGeol FGS
 FRGS
C & C Mineral Planning Services
41 South Street
Reading
Berkshire RG1 4QU

Reading (0734) 548157

London: Her Majesty's Stationery Office

1986

© Crown copyright 1987
First published 1987

ISBN 0 11 752002 0

SYNOPSIS

1 Introduction 1

2 Summary of Findings 7

RESOURCES AND RESERVES

3 Offshore Sand and Gravel Deposits 11

INDUSTRIAL FACTORS AFFECTING GROWTH

4 Economics of Production 35

5 Improvements in Dredging Technology 55

6 Future trends in Reserve Evaluation and Mining Management 61

7 The Development of Processing and Quality Control 75

INTERACTION WITHIN THE COASTAL ENVIRONMENT

8 Problems of Physical Avoidance 85

9 Shoreline Stability 91

10 Fisheries and Environmental Quality 107

WHARVES

11 Wharves and Wharf Operation 129

PROSPECTS FOR THE INDUSTRY

12 Regional Assessments 147

CONCLUSIONS AND RECOMMENDATIONS

13 Conclusions, Recommendations and Research Proposals 177

APPENDICES

1 Glossary of Terms

2 Bibliography

3 Wharf Case Studies

4 MAFF/SAGA Code of Practice for the Extraction of Marine Aggregates

SYNOPSIS

INTRODUCTION 1

1.1   TERMS OF REFERENCE OF THE STUDY

This report has been commissioned by the Department of the Environment as part of their Planning Research Programme 1984/85. The views expressed are those of the Consultants and do not necessarily reflect Government policy.

The project specification is set out below :

A.   <u>Overall Aim</u>

A.1   To provide a sound basis for assessing the potential contribution from marine-dredged sand and gravel towards meeting the future demands for aggregate minerals.

B.   <u>Specific Objectives</u>

B.1   To collect and collate existing information on the resources and licensed reserves of sand and gravel in the UK Sector of the continental shelf.

B.2   To identify and analyse

   a.   the degree to which the resources are constrained by other interests

        and

   b.   the factors which affect the operations and growth of the marine sand and gravel industry.

B.3   To define cost-effective research proposals which might reduce the constraints on the dredging industry and release additional resources for extraction.

B.4   To collect and collate existing information on wharves for marine-dredged sand and gravel in order to identify the planning issues which arise and the primary factors controlling the selection of sites.

CONTENTS :

Terms of reference of the study

   Overall Aim
   Specific Objectives

Structure of report

Administration of offshore aggregate extraction

   The Crown Estate Commissioners
   Local authorities
   Government departments

* covering :
- marine operations
- reserves & prospecting
- marine environmental problems

+ covering :
- wharf operations
- economics of production
- quality control
- planning problems

The survey extends to England, Wales and Scotland and excludes the Channel Islands, Isle of Man and Northern Ireland. The specification required an assessment of the prospects for the industry a) to 1996 and b) to 2001 on a regional and dredging ground basis as appropriate.

The project was carried out between August 1984 and July 1985 by Sea Sediments Survey & Consultancy* of Chard, Somerset in association with C & C Mineral Planning Services+ of Reading, Berkshire.

The project was guided by a Steering Committee chaired by an officer from the DoE Minerals Division. The following organisations were represented on the Committee.

Local Authorities:	Hampshire County Council	Mid-Glamorgan County Council
	Kent County Council	West Sussex County Council
Industry (BACMI):	ARC (Marine) Ltd	
	Pioneer Aggregates (UK) Ltd	Tarmac Marine Ltd
(SAGA):	Civil and Marine Ltd	
	Norwest Sand and Ballast	Ready Mixed Concrete (UK) Ltd
Government (DoE):	Construction Industry Division	Minerals Division

Administrators for owners of the seabed:
Crown Estate Commissioners (CEC)   Lewis and Duvivier (CEC Advisers)

Views expressed by the Consultants do not necessarily reflect those of individual members of this Committee.

1.2 STRUCTURE OF REPORT

The research reported in this document is directed towards fulfilling the Overall Aim of the project. The report is structured in eight parts:

- a **Synopsis** encompassing this introductory chapter together with a short chapter on the main findings of the study;

- an explanation of **Resources and Reserves** in the context of offshore sand and gravel deposits;

- an assessment of the **Industrial Factors affecting Growth** in the dredging industry;

- the identification of problems arising from **interaction** between aggregate dredging and **other interests in the coastal environment**;

- a discussion of the location and operation of **Wharves**;

- a **Regional Assessment** of the prospects for the Industry taking into account published demand forecasts, available reserves, potential reserves, the dredging fleet and the points of landing; and

- a final section setting out the main **Conclusions, Recommendations and Research Proposals** arising from this investigation.

Four **appendicies** are included to assist the reader:

Glossary of Terms, Bibliography, Wharf Case Studies, Fishing Industry Code of Practice.

A supplementary report has been placed on "open file" and can be consulted at the Library of the Department of the Environment.

That report, which is referred to in this text as 'DoE, 1986', deals with a purely factual description of the Industry as at 1985, viz:

- planning, production and administration of aggregates;

- a description of the dredging industry and background information to aggregate quality;

- dredgers and dredging methods;

- prospecting and reserve management methods.

Those topics are supported by appendices dealing with:

statistics of dredgers; a list of all wharves receiving marine aggregates; a brief discussion of aggregates from estuarine and maintenance dredging; a brief discussion of dredging worldwide; a code of practice agreed between the Crown Estate Commissioners and the Sand and Gravel Association, together with the form of a dredging licence; and statistics relating to landings of marine aggregates between 1967-84.

1.3 THE ADMINISTRATION OF OFFSHORE AGGREGATE EXTRACTION

An outline to the ownership and licensing of marine aggregate operations is presented in the following paragraphs. Further details can be found in DoE (1986).

1.3.1 The Crown Estate Commissioners

The Crown Estate owns about half of all UK foreshore, ie the area between Mean High Water and Mean Low Water Marks (except in Scotland where Spring Tides apply) and virtually all seabed to the limit of Territorial Waters. Beyond this to the edge of the United Kingdom continental shelf (ie 200m depth of water) or the median line with another coastal state, the natural resources (including mineral rights but excluding oil, coal and gas) are also vested in the Crown and under the management of the Commissioners.

The CEC licence the extraction of offshore deposits of sand and gravel under the Crown Estate Act 1961 and through the Continental Shelf Act 1964 which covers Territorial Waters outside the United Kingdom.

Two types of licence are needed: Prospecting and Production Licences; the former is a prerequisite to the latter.

In the case of applications for **Prospecting Licences** no formal consultations are initiated by the CEC although the DoE Minerals Division and the headquarters of the Ministry of Agriculture, Fisheries and Food (MAFF) or the Department of Agriculture and Fisheries for Scotland (DAFS) are informed in confidence it is intended to grant Prospecting Licences. Prospecting Licences are granted for 2 or 4 years duration and are renewable. Where a dredger is used for bulk sampling the maximum permitted tonnage that can be taken is 1000t over 2 years. When a 4 year licence is granted the total tonnage is normally increased to 2000t.

With **Production** (Dredging) **Licences** the consultation procedure is often very long because of attempts to assess the likely effects, if any, on coast erosion and fisheries interests. Applications are normally decided upon within two years, although longer periods have elapsed in contentious cases.

Since the mid 1960's the CEC have consulted Hydraulics Research on receipt of an application for a Production Licence to ensure that there is no consequential risk to the adjacent coastline (Section 9.3.3).

If the outcome of the coastal stability assessment is favourable, the CEC then ask the Minerals Division of the DoE for a 'Government View' on the application. This is obtained by consulting all Departments whose interest might be affected by the proposal, in addition to non-Government bodies directly concerned, such as British Telecom. This procedure has operated since 1970 and includes the following departments, with interests identified in brackets:

Department of the Environment* (construction materials, marine nature reserves)

Department of Transport (navigation, ports and historic wrecks)

Ministry of Agriculture, Fisheries and Food (fisheries & coast protection)

Ministry of Defence (hydrographics & lands)

Department of Energy (oil and gas pipelines)

Departments consulted as necessary include the Welsh Office and Scottish Development Department who carry out the consultations in Wales and Scotland respectively. The Department of Industry is also consulted. Other bodies consulted at this stage are the local coast protection authority (Borough/District Councils in England and Wales and Regional Councils in Scotland), and the Regional Water Authority. All these consultees seek observations from other bodies whose interest might be affected by the proposal eg the Nature Conservancy Council, Trinity House etc. In the case of the MAFF the consultation does not extend to include any balancing with land 'interests'. The views of mineral planning authorities are not sought under the present procedure.

If the 'Government View' is favourable the CEC will usually issue a licence to dredge, although it may be subject to conditions suggested by Departments as a result of the consultations. If any of the Departments object to the proposal and it seems likely that an unfavourable view is to be given, the Minerals Division informally tell CEC of this possibility so that they may contact the applicant to see whether changes/further research can be agreed in order to satisfy consultees' objections. If this cannot be achieved and the Department's objection is maintained the 'Government View' is automatically unfavourable. In these cases CEC do not issue a licence.

Of those applications submitted to DoE about half receive a favourable 'View' following initial consultations; a quarter are refused outright; and a quarter receive a favourable 'View' after initial objections have been resolved.

---

* The Construction Industry Directorate are sponsors of the aggregate industry and are responsible for economic assessment and general guidance to the building industry on public expenditure; civil engineering projects; policy on commercial investment; and monitoring of building regulations, including building materials and directives made by the EEC.

That Directorate were also responsible for co-ordinating the 'Government View' procedure until 1978. A recommendation of the Verney Committee was that this function should be separated from their role as sponsors of the aggregate industry. In the Government's Response to the Committee's findings (Circular 50/78, Welsh Office Circular 92/78) collation of the 'Government View' was transferred to the Minerals Division.

There is a special procedure for minimising the impact of dredging on sea fisheries. This has been agreed between representatives of both the fishing and marine dredging industries. Its terms were set out in the Ministry of Agriculture, Fisheries and Food publication "Code of Practice for the Extraction of Marine Aggregates" - Dec 1981 (Appendix 4).

Under the Code of Practice, which came into force on 1 January 1982, the CEC notify MAFF/DAFS HQ when either a Prospecting or Production Licence is issued. MAFF in turn send details to its Fisheries Research Laboratory at Burnham-on-Crouch, the District Inspector, and the local Sea Fisheries Committee. In Scotland, DAFS send details to its Marine Laboratory at Aberdeen, the Sea Fisheries Inspectorate, and the appropriate fishing organisations (Sections 10.1.2, 10.4.2).

1.3.2   Local authorities

Planning permission under the Town and Country Planning Acts is not required for dredging unless the site falls within the area of an administrative county. This is so if dredging takes place in the intertidal area or the county boundary extends beyond Low Water into an estuary.

In addition to the CEC licensing system, a coast protection authority may seek an Order under Section 18 of the Coast Protection Act 1949 from MAFF (Land Drainage Directorate). This enables a local authority to control the removal of material in water depths of less than 15.2 metres at LWM of ordinary spring tides in any portion of the seashore within their area or lying seaward to the limit of Territorial Waters. South Wight Borough Council have such an Order which was granted in 1951 to the former Isle of Wight Rural District Council. Dredging within those waters therefore requires both a local authority and CEC licence.

1.3.3   Government departments

A consent from the Department of Transport-DTp (Marine Directorate) under Section 34 of the Coast Protection Act 1949, as extended by Section 4 (1) of the Continental Shelf Act 1964, is also required to ensure that navigation is not obstructed or endangered. This applies to works below LWM of ordinary spring tides except, inter alia, dredging operations (including the deposit of dredged materials) authorised by any local Act. These consents are limited to 3 years and are renewable (Section 8.1.1).

SUMMARY OF FINDINGS

The UK marine sand and gravel industry currently supplies approximately 13% of the nation's demand for sand and gravel with total landings around 12Mt per annum. Some 3Mt is also landed annually at ports on the Continent. Second in the world only to the Japanese in aggregate tonnage landed from offshore sand and gravel deposits, the UK dredging industry is a leader in the design and technology of its vessels and wharves.

There are six main dredging areas. The greatest tonnages of sand and gravel are extracted on the East and South Coasts of England (approximately two-thirds of all dredged aggregate is landed at wharves in South East England). In Wales and along western coasts of England sand cargoes predominate landings, principally to wharves in South Wales. There is no dredging at present in Scottish waters.

Dredging is undertaken in all but the roughest of seas in water depths usually between 18 and 30 metres below Chart Datum. New generation vessels (two are currently under construction) will have the capability of dredging to depths of 45-50m. The majority of vessels are expected to be in operation 24 hours a day throughout most of the year.

With increasing environmental pressures over the release of new land-based reserves, attention has been focused on the importance of seabed resources to satisfy aggregate demand. The 12Mt of marine aggregate landed per annum in the UK is equivalent to some 240 hectares of land-won workings. Marine supplies can therefore be seen as reducing the impact of land extraction on agriculture, the onshore environment and amenity.

Reliance by Central Government, mineral planning authorities and the offshore sand and gravel industry generally on maintained or increased landings of marine aggregates belies the fact that little is known about the extent of resource availability offshore. Thus in the medium to longer term seabed resources are not secure to meet national needs. Total aggregate availability, both licensed and unlicensed, is uncertain due to the difficulties of recovering data in the adverse offshore environment.

This study has sought to identify and analyse the factors constraining the release of offshore sand and gravel reserves and examines ways forward which will enable the industry to fulfill its commitments and the statutory authorities to safeguard other maritime interests.

At a regional level the Consultants' greatest concern over future supplies of marine dredged aggregates relates to London and South East England. This is due mainly to the continuing demand for aggregates in London which cannot be met from indigenous land-based resources and diminishing offshore licensed reserves. There could also be repercussions on the supply of material landed in other regions of the UK.

If Government objectives regarding the future increased use of marine aggregates are to be achieved then major decisions will have to be made at national level to release substantial offshore reserves. The problem is heightened by the fact that the total quantity dredged each year has, at least since 1965, greatly exceeded the additional quantities that have been licensed by the Crown Estate Commissioners, resulting in a serious depletion of offshore licensed reserves.

The release of many dredging areas around Britain's coast has been denied primarily on grounds relating to the prevention of coast erosion and to fishing interests which it is believed unrestricted dredging could affect. More recently lengthy delays are occurring whilst evaluation of licence applications is carried out.

Fishery and coastal protection issues may often result in denial or delay of a licence because of inadequate data rather than a conviction that harmful effects will occur. Increased data generation, both site-specific and resulting from more generalised research, is seen by the Consultants as the logical step which will prevent such frustration. Similarly, more accurate and better monitored mining practices should decrease conflict between the dredging industry and other maritime interests, and should also ensure optimum recovery of valuable aggregate reserves.

A review of the 'Government View' procedure, which is currently being undertaken by the Department of the Environment, could assist in providing a better balance of views in determining licence applications. Increased involvement of mineral planning authorities in licence release decisions, with their knowledge of complementary onshore aggregate reserves, would also help in this process.

The uncertainties brought about by the lack of new licences has generally delayed further investment by many of the companies involved (for example, high investment levels are necessary for vessel replacement). This is notwithstanding the aggregate industry's commitment in principle to seek increased landings of marine aggregates.

With many inshore licensed reserves either exhausted or substantially worked out, the trend is for vessels to steam to more distant banks, thus increasing turnround time. Although many dredgers still operate on single tidal cycles (particularly along the South Coast and in the Bristol Channel), other wharves (in London and other Thames Estuary landing points especially) rely on vessels operating over two to three tidal cycles per cargo landed. This reduces the overall productivity of an operation and hence the landed cost per tonne. Escalating fuel costs since 1973 have also increased operational expenses. Vessel under-utilisation, due to decreased demand in many parts of the UK since the mid-1970's, similarly affects the economics of operations.

There appear to be no major problems with wharfage in areas where landings of marine aggregates are significant or are likely to be important in the future. Local problems do sometimes occur. These can normally be resolved by the goodwill of an operator, particularly where it is understood by the planning authority and the public that the landing of marine aggregates is important to the local economy.

The price of marine material (ex-wharf) is generally higher than land-won material (ex-pit), but proximity to markets can be decisive over which material is used. Marketing patterns relate largely to road haulage costs. Only two wharves are rail-linked (both in SE England) and little scope is foreseen for expanded sales using this method of transport.

There are no technical reasons why marine sand and gravel should not be used in substitution for sand and gravel extracted from the land. Fears over chloride and shell content of material dredged from the seabed are largely ill-founded. Widespread ignorance by potential customers and certain professions has created a 'user-prejudice' against marine aggregates which is unjustified. Better publicity is seen as a solution to this problem.

Recommendations are made for a number of research proposals which could improve overall knowledge on viable reserves, decision making processes and mining management. In view of the problems of supply facing the South East of England, priority should be focused on research projects for measures to aid the resolution of this Region's problems.

RESOURCES AND RESERVES

OFFSHORE SAND AND GRAVEL DEPOSITS

3.1   THE ORIGINS OF MARINE AGGREGATES

3.1.1   Definitions

The terms **'sand'** and **'gravel'** relate to size rather than compositon. A modified Wentworth size classification of particles is used extensively in offshore geological research, and is relied upon in this report (Table 3.1). There are slight differences in terminology and grade limits between the Wentworth scale and BS 882, the aggregate industry's standard, also shown in the Table (see also Section 7.2.1).

Description as 'sand' or 'gravel' is usually reserved for deposits consisting of naturally occurring rock fragments which have been transported by water or wind (therefore having been rounded to some degree). Sands used as aggregate usually consist predominantly of quartz, although in many deposits shell fragments may be common. The composition of gravel is often more varied; rock types in use include flint, limestone, quartzite, gneiss and porphyry.

Sand and gravel deposits are the result of long periods of weathering, erosion, transport and depositional processes. All marine deposits currently worked for aggregate were formed by processes active during the Quaternary era, a period possibly containing seventeen glaciations (Bowen, 1978). The effects of the three most recent glacial cycles are widely recognised from the British land surface; the chronology and terminology of these periods are given in Table 3.2.

A full description of the processes by which sand and gravel deposits are formed, and the distribution of Quaternary aggregate deposits on land in Britian can be found in Beaver (1968) and Archer (1972). These deposits can be broadly subdivided into:

(i)   **Valley gravels** - the channel bed deposits of rivers terraced in response to past changes in geographic conditions, including the Plateau gravels of southern England.

(ii) **Glacial gravels** - more important in 'highland' areas of Britain, although extensive deposits are worked in certain lowland areas eg East Anglia.

3

CONTENTS:

The origins of marine aggregate deposits

  Definitions
  The Quaternary legacy:
    terrestrial.
  The Quaternary legacy:
    marine.
  The significance of recent
    marine sediment transport.
  Summary: comparison with
    land-won deposits.

Regional resources and reserves
  Explanatory notes
  The South Coast
  The Dogger Bank
  The North East and Humber coasts
  Scotland
  North Irish Sea
  South Irish Sea
  The Bristol Channel and
    South West Peninsula.

TABLE 3.1 CLASSIFICATION OF GRAVEL, SAND AND FINES

BS 882	WENTWORTH-BASED
	BOULDER
	256mm—
COARSE AGGREGATE 50mm	COBBLE
	64mm—
	COARSE GRAVEL
(20mm)	16mm—
	FINE GRAVEL
———— 5mm ————	——— 4mm ———
	COARSE SAND
(1.18mm)	1mm—
FINE AGGREGATE (300um)	MEDIUM SAND
	250um—
	FINE SAND
———— 75um ————	——— 63um ———
FINES	SILT & CLAY

TABLE 3.2 BRITISH QUATERNARY CHRONOLOGY

QUATERNARY	RECENT	RECENT	0 yrs
		————	3K
		FLANDRIAN	
		————	10K
		DEVENSIAN-G	
		————	110K
		IPSWICHIAN-I	
		————	128K
	PLEISTOCENE	WOLSTONIAN-G HOXNIAN-I ANGLIAN-G EARLY & MIDDLE PLEISTOCENE	
		————	1.6M
TERTIARY		Source: Bowen 1978	

G = Glacial   I = Interglacial

Reserves are defined as explored **resources** which are capable of being worked using present day technology.

The continental shelf of the United Kingdom can be conveniently divided into eight regions in relation to bathymetry and history of deposit; some of these areas coincide approximately with arbitrary divisions used in the tabulation of Crown Estates aggregate reserve statistics and the boundaries of the latter have consequently been used in this report (Figure 3.1).

Marine aggregate resources generally, in comparison with land-won deposits, are very poorly mapped and researched. This can be attributed to:

(i) the difficulty and expense of working offshore
(ii) the problems of adequately sampling gravel
(iii) the poor return of scientific data from sand and gravel samples (cf fine sediment cores), thus discouraging academic interests.

For reasons of confidentiality, little data on the nature of commercially worked sand and gravel deposits are published or available. From consultations and literature searches the Consultants have concluded that little scientific consideration has been given to the origins of these deposits.

Current methods of prospecting sand and gravel have been investigated within this project (Section 6.2.1) and are described elsewhere (DoE,1986).

3.1.2 The Quaternary legacy: terrestrial

Glacial erosion and deposition. The approximate limits of the maximum ice cover during the final three major glaciations are shown in Figure 3.1. World sea-levels dropped in response to the increased amounts of water trapped in the polar ice caps, thus many present continental shelf areas were dry land (see below).

The passage of a glacier causes the erosion of relief features, the disruption and reworking of earlier deposits, and normally results in the widespread deposit of a uniform 'till' (stiff glacial clays, often with a boulder content). The action of meltwaters near the wasting margins of the ice sheet from localised concentrations of sands, gravels and cobbly deposits known as fluvioglacial material. These processes may occasionally form extensive areas of outwash sands eg Dogger Bank, (Oele,1971), but usually deposits are of more restricted dimensions, such as eskers and fans.

With reference to Figure 3.1 it can be seen that the North East Coast (4), Scotland (5), Northern (6) and Southern (7) Irish Sea were subject to glacial deposition during the last (Devensian) ice age, and with the exception of areas of steep slopes the bedrock is almost entirely covered with a layer of till. The Bristol Channel (8) and northern sector of the East Coast (2) were glaciated during earlier ice advances, and contain intact remnants of till as well as the debris derived from the destruction of the original glacial deposits. The South Coast (part of 8 and 1) has never been glaciated and hence contains no glacial deposits.

The development of river terraces and channels occurred throughout the Quaternary in response to complex palaeohydrological changes. Our knowledge of the stratigraphy and age of terrace remnants presently below sea-level eg outer Thames area, (D'Olier,1975) is imcomplete. Buried channels on the continental shelf however are more confidently related to periods of maximum glacial advance during which sea-levels possibly lay as low as -150m below present. The buried channels of several rivers along the South and South West coasts appear to grade to approximately -50m ODN (Mottershead,1977), and the floor of the buried Thames valley has been traced to a similar depth. The initial infill of these palaeo-valleys is often gravel, although invariably covered by thicknesses of sandy or silty overburden (Augris and Cressard,1984).

3.1.3   The Quaternary legacy: marine

Nearshore wave action has affected much of the upper continental shelf through the Quaternary as a result of relative changes in sea-level. Gravel and sand movement is intense in shallow water along exposed coasts (Section 9.2.3) and many cycles of such reworking would have been effective throughout the Pleistocene.

Although the broad patterns of worldwide sea-level change for the whole of the Quaternary have been speculated upon, detailed knowledge is only really becoming available for the last, Devensian glaciation (Figure 3.2). It can be seen that there was no single rise and fall of sea level, but a fluctuating situation where areas uncovered by ice (eg Bristol Channel, South and East Coast) within the present -15-50m ODN swathe may have been inundated up to six times. In contrast regions of till deposited during the maximum (c.20,000 yrs ago) Devensian ice advance have only been subjected to a single (the 'Flandrian') inundation by rising sea levels.

Superimposed on the worldwide sea-level changes are local variations due to landmass movement. Throughout the Quaternary unglaciated areas such as the South West Peninsula are thought to have been relatively stable. In areas covered by thick ice deposits,

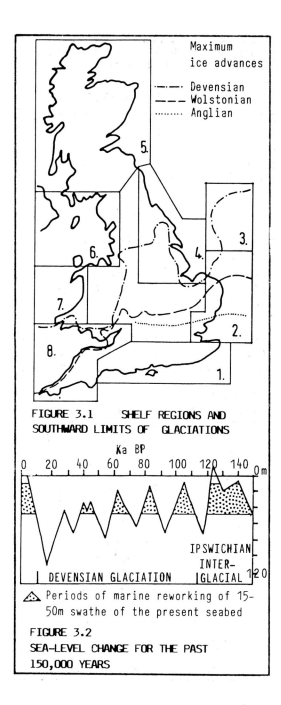

FIGURE 3.1   SHELF REGIONS AND SOUTHWARD LIMITS OF GLACIATIONS

FIGURE 3.2 SEA-LEVEL CHANGE FOR THE PAST 150,000 YEARS

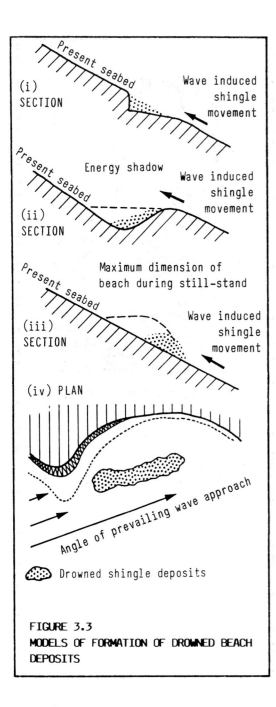

FIGURE 3.3
MODELS OF FORMATION OF DROWNED BEACH DEPOSITS

notably Scotland, the land sank under the weight of ice, with subsequent recovery. Because of the time lag between ice loading and full crustal response, many of the previous Scottish shelf areas affected by the Flandrian inundation are now tens of metres above sea-level (Cullingford and Smith,1966). Conversely, tectonic downwarping (of the Rhine graben) in the southern North Sea is constantly lowering the floor of the East Coast area (D'Olier,1975); at the present rate of lowering of around 2mm/yr deposits which were at sea-level 3000 yrs ago would now be found below 6m of water.

As well as effecting the reworking of earlier deposits and the erosion of bedrock, a transgressing (rising sea-level) shoreline also acts as an efficient sorting mill. Surf action winnows out muds and fine sands and disperses them offshore whilst trapping medium and coarse sand and shingle inside the breakers (Section 9.2). With rising sea-levels the beach sediments along a high energy shore can be transported up the continental slope. Falling sea-levels cannot reverse this process, thus many of our present day beaches on exposed coasts eg Chesil Beach (Carr and Blackley,1974) are thought to have been derived from the cumulative effects of Quaternary sea-level fluctuations.

Four situations can be envisaged where ancient beachlines are left trapped in what are now submerged areas (illustrated in Figure 3.3):

(i) Sharp steepening of the landward slope (eg resulting from earlier differential erosion of rock strata).
(ii) A reverse slope, casting a shadow zone where wave energy is ineffective.
(iii) A prolonged stand of sea-level, allowing the build-up by longshore drift of a locally extensive deposit, which is too thick to be reworked when sea-level rise is resumed.
(iv) The interference of headlands and shoal areas in reducing the energies of prevailing waves which approach the shore at an angle.

Because of their high degree of roundness, and diagnostic 'chattermarked' surfaces, shingle deposits offshore from the Isle of Wight are thought to be predominantly of 'fossil' beach origin (R A Fox, RMC, pers.comm.).

Tidal streams similar to those found today (Fiugre 3.4) can be conjectured for the high sea level stands of the Quaternary.

A tidal velocity of at least two knots (100cm/s) is required to initiate motion in 4mm gravel, and a peak surface tidal velocity in excess of three knots is probably required for substantial transport of shingle. Reference to Figure 3.4 shows that there are presently areas in the north and south Irish Sea, the inner Bristol Channel, central

English Channel, Dover Straits and Scottish Islands where tidal transport of shingle would be expected. Seabed features such as erosional furrows in gravel areas (Stride, Belderson and Kenyon,1972), and shingle waves (Hammond, Heathershaw and Langhorne,1984) have confirmed the existence of this type of movement. Bedload transport rates are undoubtedly slow, but the cumulative effects of transport during high sea levels of the past 150,000 years have been observed, with gravel layers thickening away from areas of strongest tidal flow (Larsonneur,1965; Kenyon,1970).

The **rafting** of gravel and cobbles, either by ice or seaweed attachment, is likely to have been an active process of deposit formation through the Quaternary.

3.1.4 The significance of recent marine sediment transport

Sea-level has been essentially stable over the past 3000 years, enabling establishment of the sediment circulation systems that are observed today. The importance of this transport system to aggregate resources can be usefully summarised in relation to the energy available for sediment movement.

High energy environments are confined to the shorelines of exposed coasts and areas where tidal currents exceed three knots, as discussed above. Within these areas coarse aggregate deposits (>2mm) may actively be forming and if mined, there is potential for replenishment by natural mechanisms. Apart from in estuaries, sand deposits are only usually present as ribbons on coarse substrates, and fine deposits are absent.

Moderate energy environments are zones of intense sand transport. Under tidal streams (1-3 knots) medium to coarse sand moves in bedload streams from areas of bedload parting towards sink zones (Figure 3.5). Medium sand (250-500µm particle diameter) in particular is widespread, forming sand banks, sandwave fields and extensive sand sheets often many metres thick. Shore-parallel zonations of medium and fine sand deposits occur under wave-driven energy regimes along the coast (Section 9.2). Fine deposits are generally absent.

Whereas medium and coarse sands are dredged on a limited scale for building or 'soft' sand, in most areas the demand for this almost ubiquitous sand appears to be small. Thicknesses of gravel are undoubtedly sterilised by overlying medium sand deposits, which it is not presently viable to dredge away and dump as overburden (Augris and Cressard, 1984).

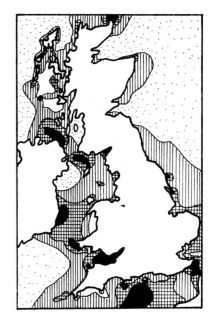

FIGURE 3.4
THE DISTRIBUTION OF PEAK SURFACE TIDAL VELOCITIES

Coarser material may remain in situ as a 'lag' deposit in areas towards the head of bedload transport pathways, where the mud and sand is winnowed away and gravel left behind. It is difficult to envisage a residual shingle of more than a few centimetres thick resulting from this process, and indeed lag deposits overlying glacial till have been reported as often little more than one cobble thick (Dobson et al,1969). Thus large areas of the seabed may superficially appear to be composed of gravel, but are of insufficient thickness to be viable as aggregate sources.

Low energy environments are sites of mud and fine sand accumulation. Gravel deposits are known to underlie overburden in several such areas (eg Maplin Sands in the inner Thames Estuary).

→ Bed load pathways
▷◁ Bed load convergences
▬ Bed load partings

Source: Kenyon & Stride, 1970

**FIGURE 3.5
SAND TRANSPORT PATHWAYS**

3.1.5   Summary: comparison with land-won materials

(i)   **Sources of aggregate.** Land-won material is worked from 'solid', 'valley' and 'glacial' deposits. The former source, where poorly consolidated pre-Quaternary rocks are crushed and screened, is not presently viable at sea. Drowned valley (river terrace and buried channel) and glacial (usually fluvioglacial) origins are recognised for material dredged offshore. The effects on these deposits of submergence and often several periods of high energy reworking, have been both positive and negative. On the one hand the gravel element of deposits such as till and weathering profiles may have been eroded, preferentially sorted and deposited in discrete workable accumulations, whereas on the other pre-existing fluvially sorted features may have been destroyed and their aggregate content spread thinly as an unworkable layer. Where intense wave and tidal action has obliterated all previous deposit morphology and produced drowned-beach or shoal deposits it is more convenient to recognise a third **Quaternary marine** source of aggregates.

(ii)   **Overburden.** Land sources of aggregate are often worked from beneath 1-2 metres of overburden. Sands and gravels have also been identified at sea with fine sediment overburdens, both deep buried valley deposits overlain by late Quaternary valley fill sequences, and thinner more extensive deposits which now lie in areas of recent mud and sand accumulation. The costs associated with present offshore mining techniques currently prohibit the working of these deposits (Section 5.2.2).

(iii)   **Deposit recognition.** Notwithstanding the problems of working underwater, it is apparent from the preceding discussion that prospecting for marine aggregate deposits involves many different considerations compared to prospecting on land. The surface sediment of the seabed often bears little relationship to the material 0.1m below the surface (for example, due to the presence of lag deposits or thin sand sheets). Offshore morphological mapping criteria will be totally different from those applicable to inland terrace remnants or fluvioglacial features, and at present appear to be largely unrecognised. Lateral variations in deposit thickness and character are also liable to be very different in marine deposits from those predictable from experience on land. Future avenues of development in offshore prospecting, directed towards coping with these problems, are discussed in Section 6.

3.2 REGIONAL RESOURCES AND RESERVES

3.2.1 Explanatory notes

In the following sections the marine geology of eight component regions of the continental shelf around the British Isles is described simplistically in relation to aggregate resources. Only areas inshore of the 50m isobath have been considered, as this is the likely maximum economic depth to which aggregate dredging could extend within the next fifteen years, the period under consideration in this project (Section 5.2.1).

For each region statistics are also presented relating to

(i) licensed reserves and
(ii) unlicensed reserves, known because they have been prospected and licences are either pending decision or have been refused. These statistics **do not** include any prospected reserves for which no Production Licence application has yet been made, but these are not thought to be extensive. The figures given for licensed and unlicensed reserves have been compiled by a Sub-Group of members from the Steering Committee for this project. The figures are the best available but are not precise and have large confidence limits (possibly $\pm 50\%$). The figures are constantly reviewed as more accurate prospecting data are generated.

The extent of areas for which Prospecting Licences have been issued (not necessarily explored) are also shown.

The maps of the following pages show generalised sediment characteristics at, but not necessarily below, the seabed. Water depths shown relate to local Chart Datum (CD), which typically lies 2-4m below Ordnance Datum. Unless otherwise indicated, the maps have been compiled specifically for this study using data from sources referred to in the text.

THE SOUTH COAST

LICENSED RESERVES: 41M tonnes

UNLICENSED RESERVES:
12 Areas refused: 25M tonnes
3 Areas pending: 60M tonnes

LICENSED RESERVES not suitable for concreting aggregates: 4M tonnes

Refusals for fisheries (9), and coast protection (1) reasons

3.2.2 The South Coast (Figure 3.6)

The shelf inside 50m depth is quite wide along this coast, varying between 16 nautical miles in Lyme Bay, 30n miles just west of Wight, 13n miles from Beachy Head to Dungeness, and narrowing at major headlands and into the Straits of Dover. The 15-35m zone typically accounts for at least two thirds of this width.

Incident wave energy decreases from west to east with the progressive shelter afforded by the coast and shoal areas from the prevailing westerly swell. The wave climate is particularly severe along westward facing coasts. Tidal energies peak south of the Isle of Wight and Portland Bill, and within the Straits of Dover (3 knots), and lie within the range of 1-2 knots along most of the rest of the coast (Figure 3.4). Strong tidal currents are also found in the western Solent (up to 4 knots).

This coast has never been glaciated, and the Quaternary largely represents a period of intense periglacial and marine erosion. In consequence a relatively thin layer of coarse 'lag' deposits overlies bedrock throughout much of the area, with thicker deposits of finer sediments only building up in sheltered areas such as the western part of Lyme Bay (Eagle et al,1978) and east and north of the Isle of Wight. Gravel transport has occurred through both shoreline transgression and, in more limited areas, tidal scour. Extensive gravel sheets and banks were formed by these processes, both intertidally (eg Chesil and Hurst beaches) and subtidally (eg gravel banks of the west Solent and Needles area, gravel sheets south of the Isle of Wight).

Two major early or pre-Quaternary sources can be identified for these gravels. The first is Tertiary weathering and erosion of the Chalk outcrop, which would have produced a mantle of flint debris throughout the area. Secondly, the erosion of terraces of the former Solent river (which once drained the south western areas of the Hampshire Basin) would have yielded large quantities of gravel in the Solent area. Some remnants of these terraces undoubtedly remain intact (Dyer,1985) and are possibly the source of gravels dredged within the eastern Solent and Owers Bank grounds.

Present day tidal processes, possibly active through high level sea stands of the Quaternary, are carrying sand both east and west from the bed-load parting area south of the Isle of Wight (Figure 3.5, from Stride et al,1972). Throughout much of the area west of the Island sand transport is represented only by thin sand ribbons on the gravel/rock basement. Between Wight and the Dover Straits there are several areas of continuous sand cover/sand waves, burying any underlying gravels.

The area has been extensively prospected, and licence applications are pending in areas

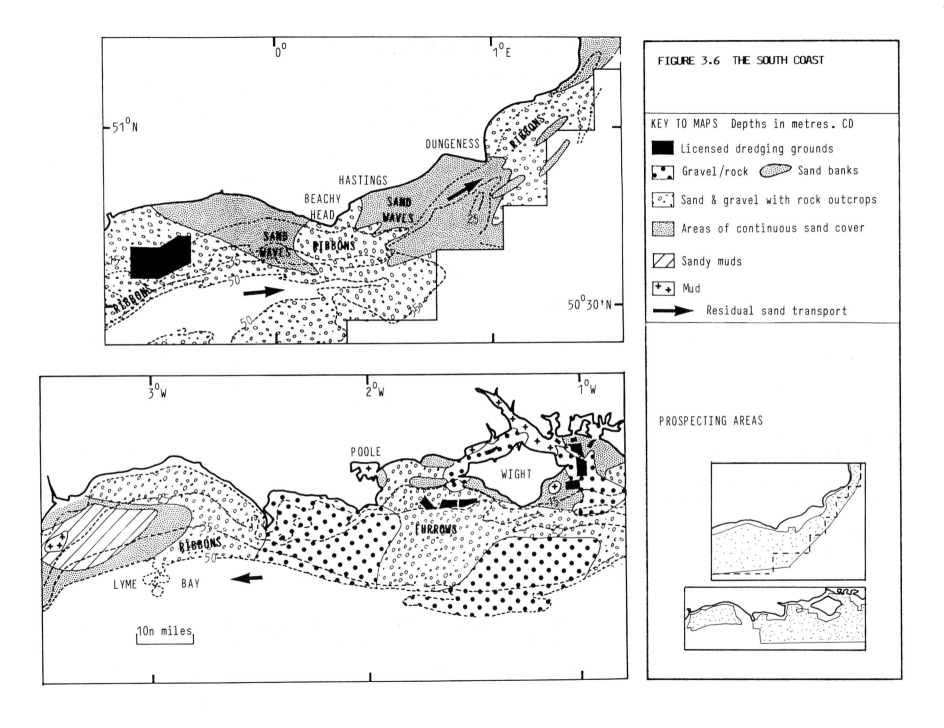

FIGURE 3.6 THE SOUTH COAST

EAST COAST

LICENSED RESERVES: 41M tonnes

UNLICENSED RESERVES:
5 Areas refused: 18M tonnes
4 Areas pending: Undisclosed

LICENSED RESERVES not suitable for concreting aggregates: 5M tonnes

Refusals for fisheries (3), navigation (1) and coast protection (1) reasons

THAMES ESTUARY

LICENSED RESERVES: 15M tonnes

UNLICENSED RESERVES:
1 area refused: Undisclosed
1 area pending:

LICENSED RESERVES not suitable for concreting aggregates: 15M tonnes

Refusal for navigation reasons

---

south-east of Wight and south of Hastings. There are wide areas between -25m and -35m CD, largely beyond the reach of the present aggregate fleet, which may in the Consultants' view be important resource areas.

3.2.3   The East Coast and Thames Estuary (Figure 3.7)

The Southern Bight of the North Sea is a shallow area, with almost all gravel lying inside the 35m isobath. Large areas of the Thames Estuary are shallower than 15m.

Wave energy is moderate to low, increasing into the less sheltered areas north of the Norfolk Banks. Tidal currents do not exceed 3 knots, except in the North Foreland area, and decrease north of the Anglian coast to 1-2 knots (Figure 3.4).

The Quaternary history of the area is complex. Channel deposition within the proto-Thames, from at least as early as the beginning of the Quaternary, has left a complicated series of channel floor and terrace remnants in the outer Thames Estuary (D'Olier,1975) and along the East Anglian coast to the latitude of Norwich (Beaver,1968). Ice cover extended to affect all areas north of the Stour/Orwell/Colne rivers, although the southward penetration of ice decreased through each of the last three glaciations (Figure 3.1) so that only areas north of Norfolk were affected by the two final (Wolstonian and Devensian) ice advances. Consequently glacial deposits in the offshore area south of Great Yarmouth were reworked by marine and periglacial process during the Devensian and Wolstonian periods.

The gravels thoughout the area are thought to be largely derived from the Tertiary rivers, Veenstra (1969) identifying a flint-limestone-quartzite gravel suite. The coarse deposits of the Cross Sands area are less well studied and may also contain material of fluvioglacial origin.

North-west of Great Yarmouth, within the areas covered by Devensian glaciers, a layer of till mantles the seabed. A very thin lag deposit of gravels and boulders is found at the seabed as a result of erosive action during the Flandrian transgression. This layer may thicken in excess of 1m in localised areas, possibly reflecting the presence of fluvioglacial materials. Gravels in this area are probably of Veenstra's (1969) sandstone-limestone-porphyry type, derived from Northern England.

Aggregate deposits are worked between Great Yarmouth and the outer Thames Estuary. The distribution of deposits has been controlled in two basic ways by late Quaternary and recent processes.

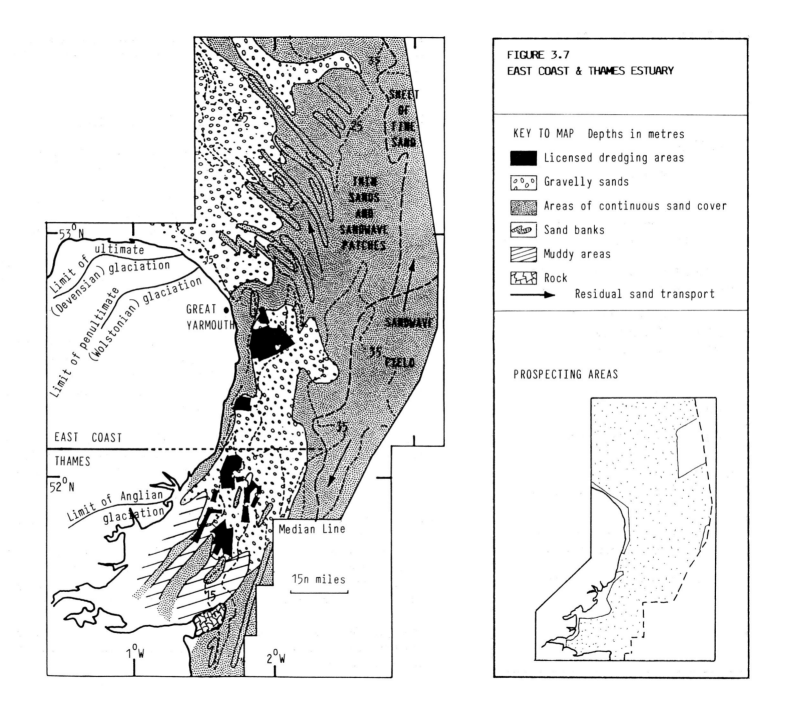

* Excess sand within aggregate deposits is a commercial problem throughout the East Coast and Thames Estuary.

THE DOGGER BANK

NO LICENSED OR UNLICENSED RESERVES

Firstly tidal streams have effectively shaped a series of shoals, sandwave fields and sheets of medium and fine sands (Houbolt,1968; McCave,1971) which bury any underlying gravel strata across wide areas of the southern North Sea (the large amounts of sand* present reflects the enormous fluvioglacial input of the Quaternary; Kruit,1963).

Secondly, considering the moderate tide and wave energies available, it is probable that many of the gravel deposits of the East Anglian coast have been 'let down' rather than transported to their present positions. Thus although many of the gravels of this area are known to have been subject to periods of beach reworking (deduced from pebble roundness and chattermarkings), most have probably undergone minimal transport, and may encompass intact remnants of river and fluvioglacial terraces. The wide variety of origin of known aggregate deposits is reflected in the geometry of the extraction grounds, some being of limited extent but considerable thickness, whereas others cover a large area but are only a few metres thick, overlying clay or chalk.

The region has been intensively prospected, and five Production Licence applications are under consideration.

3.2.4 The Dogger Bank (Figure 3.8)

The Dogger Bank contains approximately 400n miles^2 of seabed between -10 and -35m. The Bank is very exposed to waves from the northerly sector, and is an area of moderate to low tidal energies (0.5-2 knots).

The Bank is thought to have been the southern limit of Devensian ice in this part of the North Sea (McCave et al,1977) and is an accumulation of fluvioglacial outwash sands overlain by Recent sands. Areas of gravelly sands have been reported from the Bank, of the sandstone-limestone-porphyry type (Veenstra,1969). Fine sands and muds floor the Outer Silver Pit (Eisma,1975) and a sandwave field and relict sandbank system occupy the south-west corner of the Bank.

No gravel has been won from the Dogger Bank area, and although Prospecting Licences have been issued for areas north and south of the Outer Silver Pit, no Production Licence applications have been submitted. In view of the limited amount of available data on the geology of the Bank it is difficult to review its aggregate resource potential.

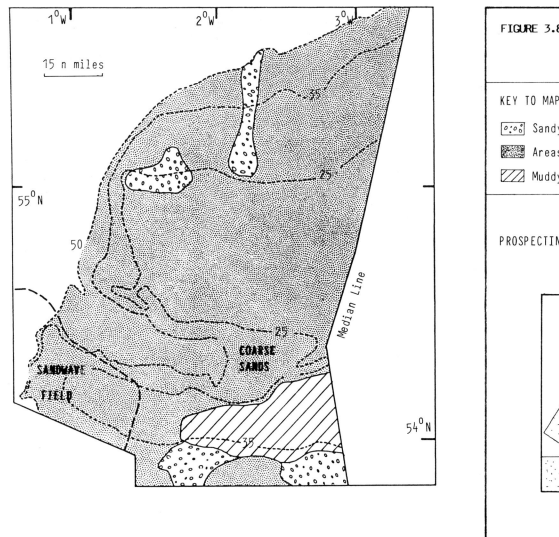

> **HUMBER**
>
> LICENSED RESERVES: 32M tonnes
>
> UNLICENSED RESERVES:
> 11 Areas refused: 30M tonnes
> Nil areas pending
>
> LICENSED RESERVES not suitable for concreting aggregates: 2M tonnes
>
> **Refusals for fisheries (10) and coast protection (1) reasons**

3.2.5 The East and Humber coasts (Figure 3.9)

South of Flamborough Head the shelf area inside the 50m isobath is very wide, often in excess of 50 nautical miles, with the 15-35m swathe occupying about 50% of this width. North of Flamborough the submarine slope becomes very steep and the 50m isobath normally lies 5-10n miles from the coast.

Wave energy is severe along the North East Coast, moderating into shallower waters south of Flamborough. Tidal stream velocities are strongest inshore around the mouth of the Humber (2-3 knots) and decrease northwards and offshore, peak velocities being less than 1.5 knots throughout much of the area (Figure 3.4).

All of this area was covered by ice during the maximum of the Devensian glaciation, and a mantle of till is ubiquitous south of Flamborough Head. North of this point increased exposure to wave action and the steepness of the coastal slope has resulted in much of the till being eroded away by the transgressing Flandrian seas, leaving a zone of rock outcrop and localised residual sands and gravels (Dingle,1970). Off the mouth of the Humber the surface of the till is composed of a thin gravel lag deposit produced in situ by the winnowing action of the sea on the glacial boulder clays. These gravels thicken in the area encompassing the Inner Silver Pit, where the Production Licensed areas are grouped. The increased gravel thickness in this region may be attributable to the presence of fluvioglacial deposits. The thickness of the worked layer is normally only of the order of one to a few metres. Thickest deposits are found around the New Sand Hole area, where anchor dredging is practised. The New Sand Hole channel is thought possibly to have formed under tidal action at a period when the entrance to the Humber Estuary was well seaward of the present, receding coast (Donovan,1973), and gravel deposits in this locality may well be the product of tidal scour and accumulation.

Tidal transport during the Recent period has formed an extensive sandwave field along the eastern limits of the Humber area, a series of sand shoals south-east of the Inner Silver Pit (eg Dowsing Shoals) and a tongue of medium sand sheets extending southwards from the Farne Islands. Mud deposits have accumulated off the mouth of the River Tyne in response to the high suspended sediment discharge of the river. Man has influenced the distribution of gravel deposits along the coast between Blyth and Sunderland through many years of dumping colliery waste (unsuitable for concreting aggregates).

Prospecting Licences have been issued for most of the seabed areas off the Humber and North East Coasts. Gravel prospects are known to be excellent off the mounth of the Humber, some 30 million tonnes having been identified although Production Licences have

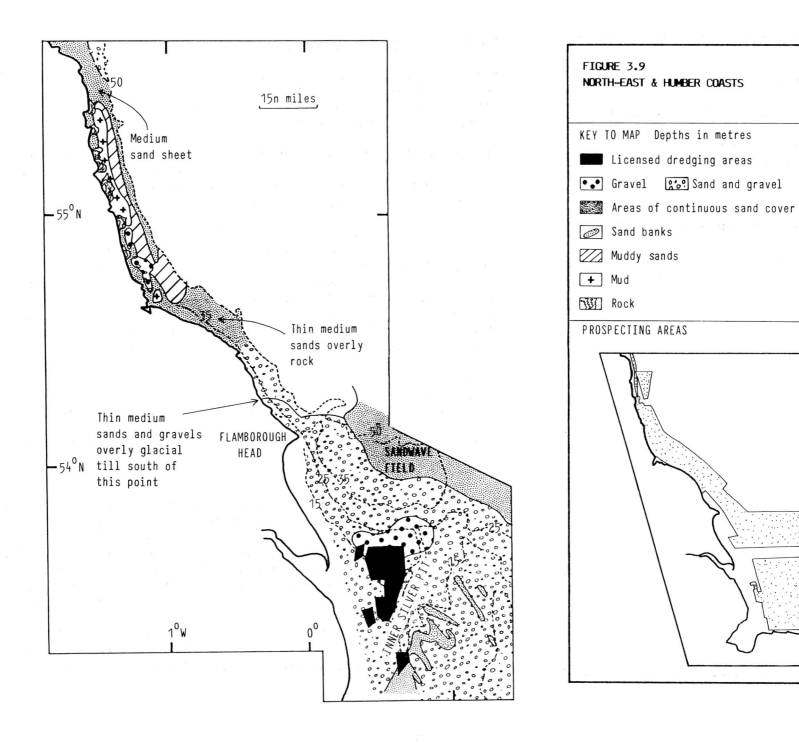

> **SCOTLAND**
>
> **NO LICENSED RESERVES:**
>   3 Areas rejected: 11M tonnes
>
> **Refusals for fisheries reasons**

been refused on fisheries grounds. No Production Licences have been applied for north of Flamborough Head, and gravel reserves are probably limited to minor pockets, and if of local provenance may also be of poor stone quality.

3.2.6 Scotland (Figure 3.10)

Three sub-regions of Scotland's long and intricate coast probably contain the greatest potential for aggregate reserves, if only because of local widening of the shallow zones of the continental shelf.

The Firth of Clyde has been mapped by BGS with particular reference to the occurrence of aggregate deposits (Deegan et al,1973). Results were not encouraging, surface gravels only being encountered on beaches and in the very shallow sublittoral zone. Deposits are of restricted dimensions, and usually of local provenance thus the stone content is liable to shrinkage problems (DoE, 1986). Very sandy gravel reserves were identified in the Ayr-Irvine Bay area.

The Firth of Forth prospects are limited to buried channel gravels with overburden problems, and lag gravels veneering the offshore till deposits of the Wee Bankie, Cockenzie Reef, Marr Bank, Bell Rock Bank and Scalp Bank (all 30-45m depth). All stations cored in the latter areas showed the gravel residual deposits to be less than 0.3m thick (BGS, Thompson, 1973).

The Moray Firth. Good quality gravels in shallow waters have been reported from reconnaissance surveys in this area (Eden,1970), but the results of subsequent investigations are not known.

Eden (1970), in considering marine gravel prospects in Scottish Waters, concluded that:

> "Widespread Marine gravel deposits similar to those off the southern coasts of England are not expected and do not appear to occur since in general the relics of earlier Scottish beaches and sub-aerial deposits lie above sea level rather than below it." (As a result of isostatic emergence - Section 3.1.3).

These conclusions appear to have been borne out by reports of disappointing prospecting results and very limited Production Licence applications during the subsequent fifteen years, although there still appears to be considerable interest in the region.

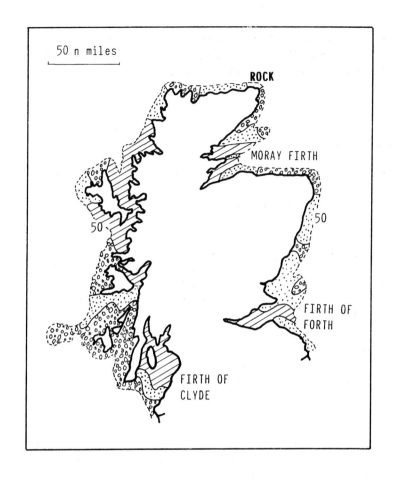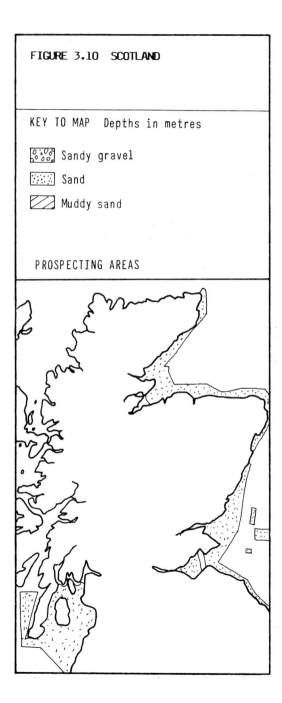

FIGURE 3.10 SCOTLAND

NORTH IRISH SEA

LICENSED RESERVES:
- Sand: 10M tonnes
- Gravel: 0.5M tonnes

UNLICENSED RESERVES:
- Areas refused: 2M tonnes
- Areas pending: 10M tonnes

LICENSED RESERVES not suitable for concreting aggregates: 2M tonnes

Refusal for coast protection reasons

SOUTH IRISH SEA

NO LICENSED OR UNLICENSED RESERVES

3.2.7 North Irish Sea (Figure 3.11)

The Irish Sea east of a line joining the Mull of Galloway and Anglesey is shallower than 50m. Between one half and two thirds of the 7000n mile^2 of sea area lies between 15 and 35m water depth. Tidal currents attain values of 1-2 knots throughout the Sea, only reaching velocities in excess of 3 knots in a localised zone extending north-east of the Isle of Man (Figure 3.4). Moderate wave energies prevail.

The region was completely covered by ice during the maximum advances of the Devensian glaciation, forming extensive till deposits, and a complex of bedded proglacial sediments is known to have been laid down south and north-east of the Isle of Man during deglaciation (Pantin, 1977 & 1978). Where till underlies the seabed a ubiquitous coarse lag deposit has formed during Flandrian and Recent times. This residual layer thickens in certain areas to form gravelly deposits in excess of 1m depth.

An area of mud accumulation has formed along the Cumbrian coast in response to transport patterns imposed during the Flandrian transgression (Williams et al, 1982). This mud area is encompassed by an active zone of sand transport, with continuous sand cover thinning offshore into sand ribbons and isolated sandwaves. More than 50% of the 15-35m seabed area is occupied by mud and medium sand deposit.

The aggregate resources of the northern Irish Sea are thought to be the result of local marine concentration of till-derived lag deposits, probably enhanced by the occurrence of fluvioglacial sands and gravels. Few of the known gravel areas have been prospected, probably because most of the deposits of favourable appearance are distant from the Mersey and lie in 25-35m depths, beyond the range of smaller dredgers. Coarse gritty sand with a minor gravel component is worked off the North Wales coast, and building sand is dredged at the entrance to the Mersey. Further gravelly sand reserves have been identified in the vicinity of the North Wales ground.

3.2.8 South Irish Sea (Figure 3.12)

The southern Irish Sea inshore of the 50m isobath consists of Caernarvon and Cardigan Bays, varying in width from 2-30 nautical miles. The energies of the prevailing wave climate are moderate, and tidal streams vary from 3 knots in the restricted vicinity of St.Davids Head to less than one knot along the Cardigan shore.

Both bays were glaciated during the Devensian, leaving a continuous mantle of till throughout the area. The interaction of Irish Sea and local Welsh ice is thought to be

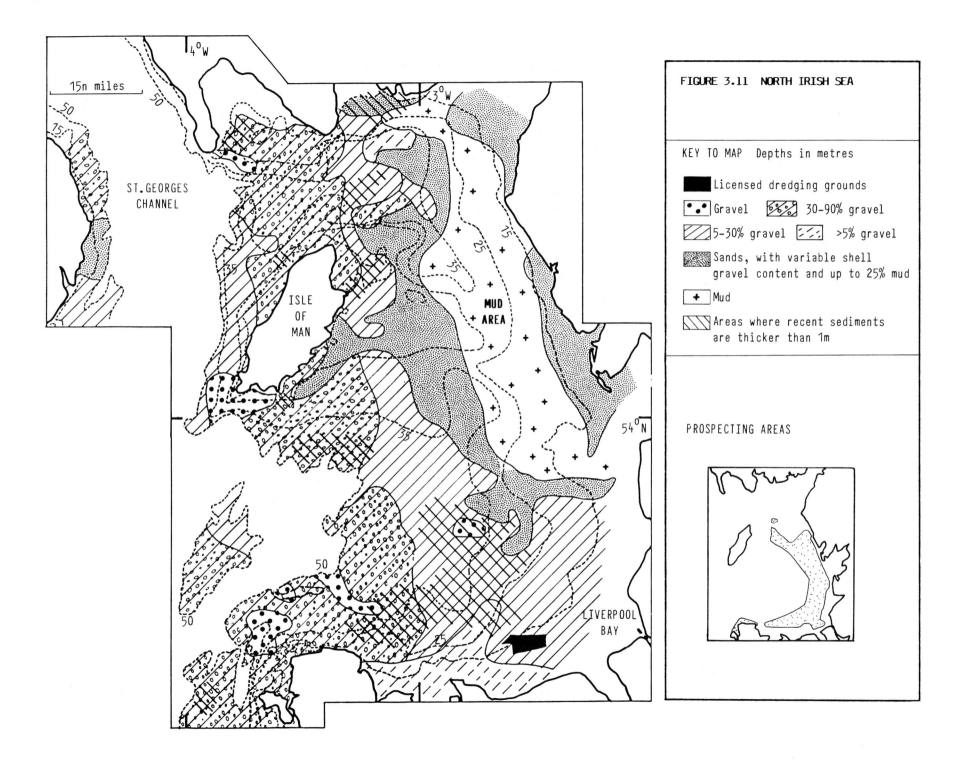

FIGURE 3.11 NORTH IRISH SEA

responsible for several large moraine features along the north-east shores of Cardigan Bay (Garrard,1977). A coarse lag deposit of cobbles and shell gravel veneers the till layer, and is observed in the few core samples taken to be typically 'one cobble thick' (Dobson et al,1971).

Sand transport is active in the outer and central areas of the bays under the present tidal regime and an extensive sand sheet occupies deeper water, the periphery of which is seen as sandwave trains along the north-west margins of Cardigan Bay. Finer sands and muds have accumulated along the shallower eastern side of the bay.

No aggregate prospecting or production has taken place in the region.

3.2.9 The Bristol Channel and South West Peninsula (Figures 3.13 and 3.14)

Some 45 nautical miles separate the Welsh and North Devon coasts along the 50m isobath of the outer Bristol Channel. The width of the area inshore of 50m depth rarely exceeds 10n miles on the north Cornish coast, and on the south Cornish and Devon coasts varies between 2 and 6n miles.

High wave energies prevail except within the more sheltered inner Bristol Channel area. Conversely tidal currents are strongest in the inner Bristol Channel, attaining peak velocities in excess of 3 knots. Away from the Bristol Channel spring tide velocities rarely exceed 2 knots, and off the south Cornish coast fall below one knot.

The outer and central areas of the Bristol Channel and the north Cornish coast are thought to have been under ice during the penultimate (Wolstonian) glaciation, but not during the Devensian (Figure 3.1). There is no evidence to suggest that either the south Cornish coast or the innermost Bristol Channel were ever glaciated. The intenseness of wave action along the Pembroke and north Cornish coasts during the ensuing tens of thousands of years has obliterated all submerged traces of till, the coarse content of the glacial deposits presumably remaining as a component of local gravel sheets. Similarly the powerful tidal currents of the central Bristol Channel area have scoured the Channel floor down to bedrock and boulders. Till, overlain by up to several metres of lag and fluvioglacial gravels, has been recognised in the outer Swansea Bay and Scarweather Sands area (Blackley,1978) and possibly remains beneath the lag gravel, cobble and boulder deposits which lie peripheral to the scoured rock area between the Welsh and north Devon coastlines (Murray et al,1980).

BRISTOL CHANNEL

LICENSED RESERVES: 40M tonnes

UNLICENSED RESERVES:

3 Areas refused: 5M tonnes
Nil areas pending

NIL RESERVES not suitable for concreting aggregates.

Refusals for navigation reasons

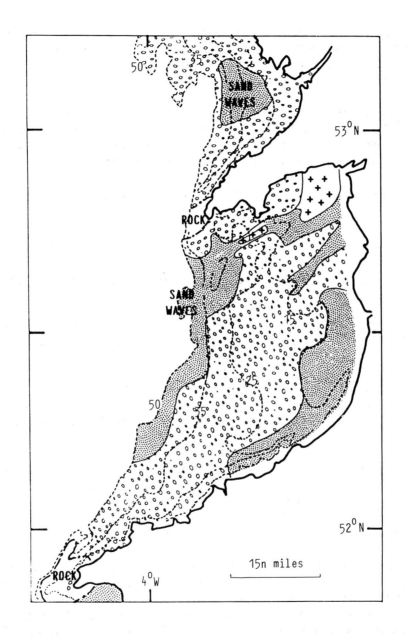

FIGURE 3.12 SOUTH IRISH SEA

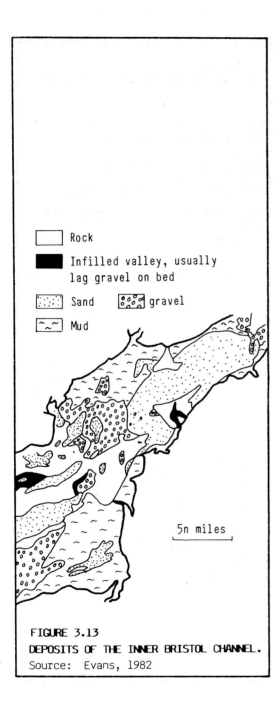

FIGURE 3.13
DEPOSITS OF THE INNER BRISTOL CHANNEL.
Source: Evans, 1982

Flandrian and present day tidal and wave energies have fashioned an extensive zone of sand transport in the outer Bristol Channel, comprising sand sheets, sandwave fields, areas of sand ribbons and a complex of large sand shoals to the south and south-east of Swansea Bay. Similarly the highly dynamic inner Bristol Channel contains thicknesses of Recent deposits varying from coarse gritty sand to medium sand shoals, and areas of settled and fluid muds (Kirby and Parker,1983; Evans,1983). Active gravel waves have been observed in the strong tidal areas of the central Bristol Channel, attesting to the existence of moderate thicknesses of fine gravels. The north Cornish coast contains a sequence of inshore fine sand deposits, rock outcrops encompassed by coarse shelly sands and gravels and deeper water medium sand sheets, all thought to be the result of present day processes and all largely unmapped. Recent sand deposits mantle much of the south Cornish coast, and there are equally large areas of exposed bedrock. Wave-formed sand and gravel deposits are known in 35-50m depths off Plymouth (Flemming and Stride,1967; Eagle et al,1979).

Production Licences are currently restricted to the central and inner Bristol Channel areas where coarse gritty sand is exclusively worked. Much of the Bristol Channel and south Cornish coast has been the subject of Prospecting Licence application, but no applications have been made for the licensing of new areas.

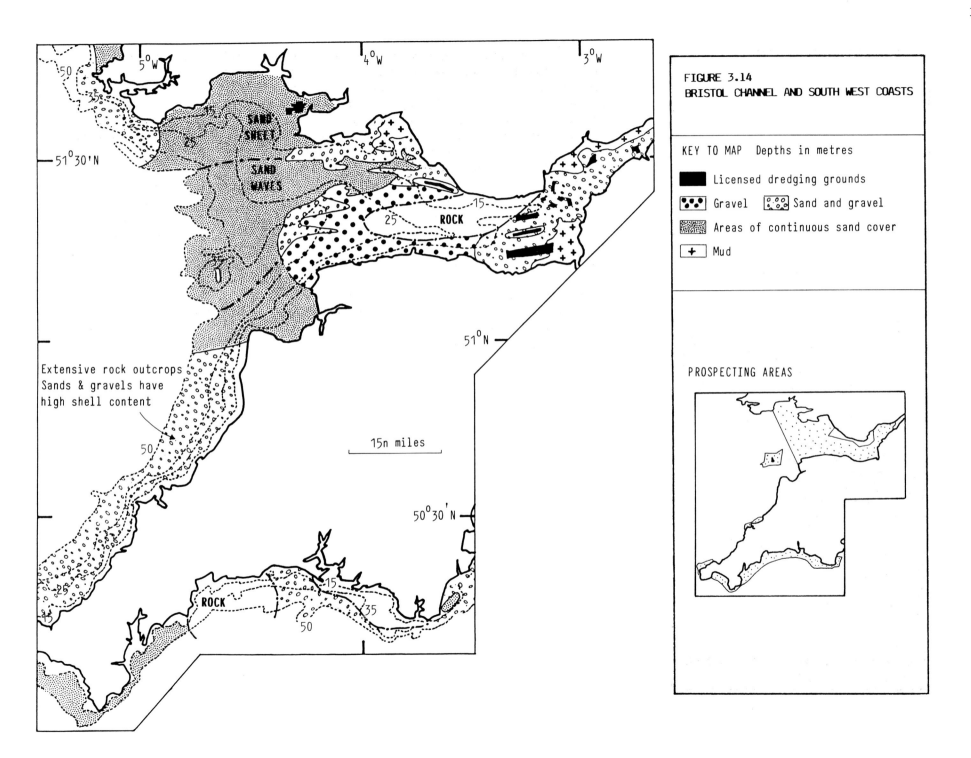

INDUSTRIAL FACTORS AFFECTING GROWTH

ECONOMICS OF PRODUCTION

4

CONTENTS:

Production statistics

    National production
    Regional production and reserves

Production costs

    Royalty rates and rents
    Dredgers and dredging costs
    Wharf and processing costs
    Price structures

Grant aid

    Ships
    Waterways
    Rail links
    Other aid

Market influences

    Inter and intra company policies

4.1 PRODUCTION STATISTICS

4.1.1 National Production

The historical production of aggregates is described in DoE, 1986. Unlike land-won sand and gravel, the production of marine sand and gravel has to be considered by reference to both where it is extracted and landed since the two can be separated by over 160 km. The dredging areas and the 1984 markets are shown on Figures 4.1 A and B.

The bulk of marine sand and gravel is dredged off the coast of England with small amounts off the Welsh coast. No offshore dredging has yet taken place in Scottish waters for the aggregates market.

The total quantities dredged, including exports to the Continent, rose to a peak of 17.42 Mt in 1979. Current rates of extraction are between 15-16½ Mt per annum.

In 1984, of the 15.5 Mt dredged 11.1 Mt (72%) was landed in England, 1.6 Mt (10%) was landed in Wales and 2.8 Mt (18%) was landed on the Continent. During the period 1968-84 the respective share of the market has varied only slightly with an annual average of 67.1% for England, 10.7% for Wales and 22.2% for exports. With an average landing of 12 Mt per annum to UK wharves this is equivalent to some 240 hectares of land-won workings. Marine supplies can therefore be seen as reducing the impact of land extraction on the environment.

In addition to the above quantities which are dredged under aggregate licences ie material primarily for use in concrete, the CEC also grant short-term licences for bulk fill mainly for use in harbour works (eg Goodwin Sands to Dover Harbour) or marshland reclamation (eg Thames Estuary to the Stone Marshes, Kent). Table 4.1 differentiates between these two types of licensed operations in terms of the total quantities dredged annually (Section 4.2.1).

4.1.2 Regional Production and Reserves

Figure 4.2 illustrates production from each of the main sea areas where dredging takes place. From this it can be seen clearly that during the 1970's East Coast production

FIGURE 4.1A
PRINCIPAL DREDGING AREAS
Scale 1:2,000,000

Landings in 1984   (> 10,000t)

LIVERPOOL BAY:
    Lancashire   137,385
    Merseyside   329,916
    North Wales   60,839

BRISTOL CHANNEL:
    Avon   376,507
    Devon   119,292
    Gwent   600,500
    Somerset   76,563
    S Glamorgan   416,265
    W Glamorgan   501,270

SOUTH COAST:
    Continent   114,418
    Dorset   133,197
    E Sussex   1,368,092
    Greater London   34,743
    Hampshire   1,670,849
    Isle of Wight   110,619
    Kent   308,298
    Suffolk   66,020
    W Sussex   316,879

Note: Gravel resources in licensed areas are not uniformly distributed.

Source: DoE/CEC

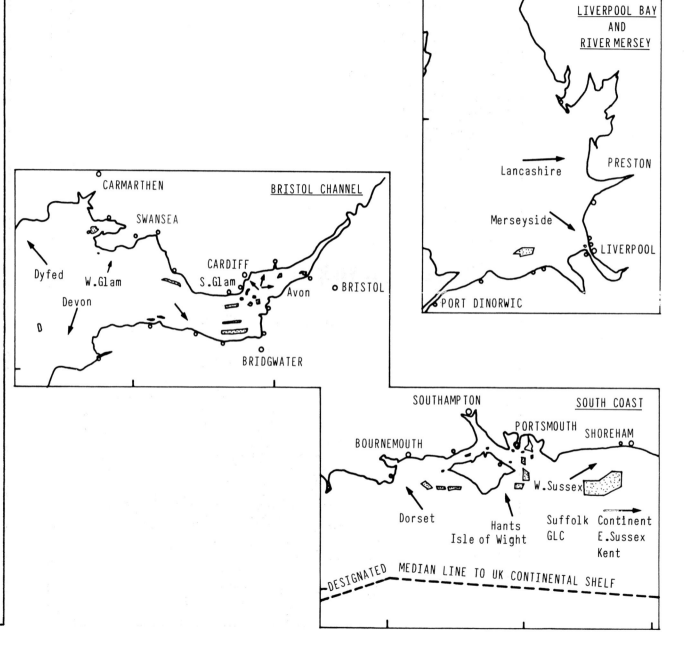

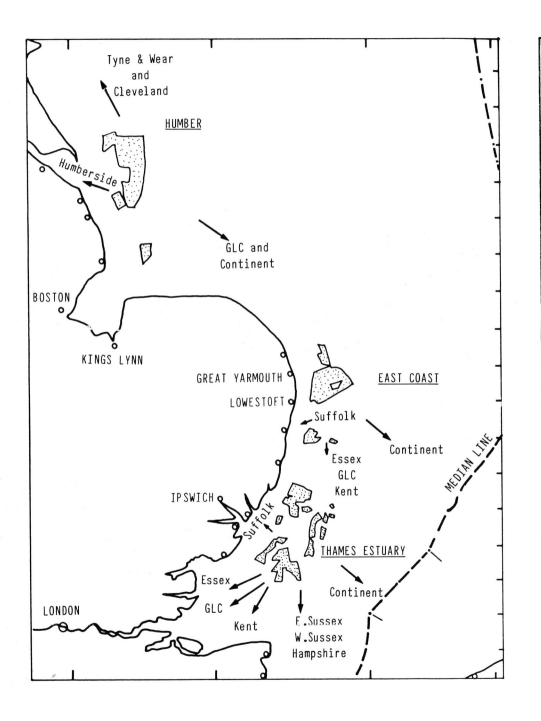

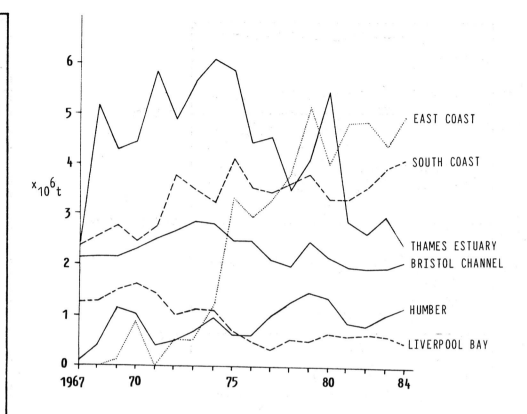

FIGURE 4.2 PRODUCTION FROM PRINCIPAL SEA AREAS

Note: 1975 Figures relate to a 15 month period.
Source: DoE/CEC

increased dramatically to become the most important dredging ground on the UK continental shelf whereas Liverpool Bay and the Thames Estuary showed a decline over the same period. All of the other areas have had fairly steady rates of production.

Figure 4.3 shows graphically landings for each of the Aggregate Working Party areas including small amounts to the Isles of Scilly, Channel Islands and to Scotland. Full statistical data are given in DoE, 1986. By far the most important area for landings is South East England with significant amounts also landed in South Wales, the South West and Northern regions.

Figures for **licensed reserves** held by individual companies, especially specific banks, are confidential. The data presented in Section 3.2 are a generalised and imprecise assessment (see caveat in Section 3.2.1) of UK reserves as at May 1985 made by the Industry in consultation with the CEC. It should be noted that significant reserves

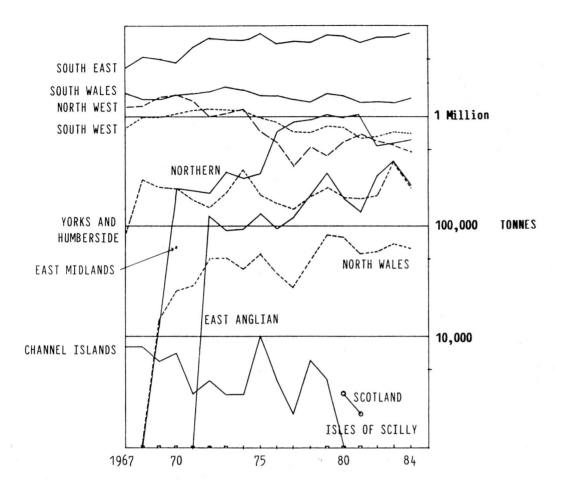

FIGURE 4.3
SEMI-LOG PLOT OF MARINE AGGREGATE LANDINGS BY ECONOMIC PLANNING REGIONS

Notes:
i) excludes extraction from rivers

ii) this is a semi-log plot and cannot be scaled.

Source: DoE

off the South Coast are awaiting a 'Government View' (ie applications not yet determined) and that dredging licences for reserves in Scotland (Section 3.2.6), the Humber (Section 3.2.5), East Coast (Section 3.2.3) and South Coast areas (Section 3.2.2) have been refused.

The total annual tonnage which could theoretically be dredged varies from year to year depending on the granting of new licences by the CEC and the surrender of old licences by the Industry. In recent years this figure has been around 36-41 Mt (including small amounts for use as fill). With annual production around 15-16½ Mt it is clear that whilst only some 40% of the licensed quantities are taken each year,

Year	Quantity extracted from Licences (Tonnes)	Fill (Tonnes)	Total royalties including Dead Rent £
1965	7,176,796)	Any fill not	196,000
1966	7,225,944)	separately	197,000
1967	8,433,971)	recorded	226,000
1968	11,701,395)		321,000
1969	12,631,872)		383,000
1970	12,840,620)		392,000
1971	13,140,341)		365,000
1972	13,653,618	3,129,369	554,000
1973	14,627,030	6,578,743	764,000
1974	15,494,928	3,269,966	926,000
1975	17,284,047	2,954,126	973,000
1976	14,943,465	1,493,047	992,000
1977	14,826,334	282,657	1,206,000
1978	15,006,421	857,426	1,184,000
1979	17,422,591	883,965	1,599,000
1980	16,927,878	870,838	1,697,000
1981	14,609,491	1,306,646	1,786,000
1982	14,532,583	2,091,305	2,287,000
1983	15,182,931	1,402,989	2,284,000
1984	15,460,035	1,220,098	2,580,000

TABLE 4.1   QUANTITIES & ROYALTIES

Notes:

i) Quantities dredged apply to a calendar year and include exports to the Continent. Royalties are taken on financial year to 31st March.

ii) 1975 dredged figures relate to a 15-month period.

Source: CEC

a) the industry could take more but the offshore reserves would be depleted more rapidly unless new licences to dredge comparable reserves were forthcoming, and

b) a proportion of the total licensed quantity may no longer be available to the Industry because of the exhaustion of banks. The Industry have not chosen to surrender those licences at this stage, either because the remaining deposits might be utilised as bulk fill material or possibly that a part of the reserves lie in water too deep to be dredged by vessels currently working those banks. The total quantity of the former was estimated to be about 28.0 Mt at May 1985.

The Consultants see this problem as a reflection of the serious lack of licensed reserves. In such an uncertain situation it would be commercially irrational to significantly increase the quantities taken under licence without there being adequate licensed reserves upon which the industry could sustain any increased level of production.

Total licensed reserves unworked and suitable for concrete aggregates at May 1985 were estimated by the CEC to be around 189.5 Mt but it is very important to note that the **life expectancy of dredging grounds CANNOT** be deduced by the production statistics given above and shown in Figure 4.2 since there are several factors involved which distort the true availability of licensed reserves:

(i) Probably the most important factors, which also apply to onshore operations, are the geographical distribution of reserves and by which companies they are held. Some companies may only have very small reserves in an area whereas another company could hold the major share. The size and extent of remaining reserves will be partly dependent on what new tonnages are licensed (Figure 4.4).

(ii) The ability, in terms of dredger capacity and movements, and the operating policies of individual companies to rest or conserve banks in specific areas. Company attitudes can be influenced by changes in market patterns and in a few cases by agreement with the CEC to reduce production for coast protection monitoring.

(iii) No distinction is made in Section 3.2 between the availability of specific grades of material, other than that which is not considered suitable for concreting aggregates; for example, whether sufficient coarse aggregate exists within a licensed area to make dredging of all of the remaining reserves economically viable.

Thus, **what overall might appear to be a generally satisfactory situation in terms of licensed reserves for individual sea areas** (when viewed in relation to the data in Section 3.2 and the production shown in Figure 4.2), **distorts the true picture for individual companies. Because of the uneven distribution of reserves between licensed areas and between different operators, some companies are experiencing difficulty in their forward planning.** Hence the need to submit further licence applications in an attempt to maintain continuity of production. In some instances, such as in Scottish waters (Section 3.2.6), applications have been submitted in an attempt to increase marine landings overall by trying to enter new market areas.

Figure 4.4 shows the total quantities dredged from the UK Sector of the continental shelf from 1965 to 1984 compared with the additional quantities licensed each year by the CEC. This clearly indicates that **recent annual rates of extraction are around 13Mt in excess of that quantity granted in new licences each year. The overall effect of this is that the level of licensed reserves has steadily decreased over the period.**

4.2 PRODUCTION COSTS

4.2.1 Royalty Rates and Rents

Royalty rates are variable costs which are reviewed annually (previously every three years but due to rapid inflation in the 1970's the Crown were obliged to reassess the frequency of reviews). Royalties are payable twice annually and licensed operators have to pay a **dead rent** which reflects 20% of the total quantity allowed to be dredged each year; dead rents are deductable from royalties. This is to ensure that there is an incentive for operators to dredge at least 20% of the licensed amounts from each area every year. Both royalties and dead rents are linked to the Retail Prices Index (RPI).

Table 4.1 shows the quantities extracted from dredging licences for the years 1967-84 and the total royalties paid. The royalties include the sums received for reclamation material taken from licences issued specifically for fill contracts. The quantities taken from "fill" licences are also shown on the table and are additional to the ordinary licences.

Royalties for well established licences are generally the same in each sea area. However, this does not imply that a new licence would necessarily have a royalty at the existing rate.

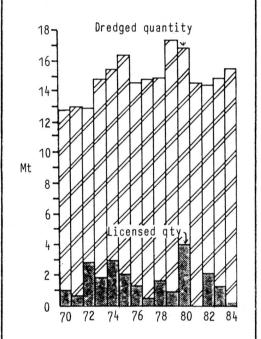

FIGURE 4.4
QUANTITIES DREDGED
COMPARED WITH NEW QUANTITIES LICENSED

Note : 1975 dredged figure relates to a 15 month period

Source : DoE and CEC

Typical dredgers from the UK fleet

Several criteria are used by the CEC to assess an appropriate royalty. These are:

(i) To obtain the best consideration for the Crown taking into account relevant circumstances but excluding monopoly value.

(ii) The type and quality of the material.

(iii) The market to be supplied and competition from other sources, including land pit and quarry prices.

(iv) The production costs.

(v) The selling price.

(vi) The profit to the dredging companies.

(vii) General movement of selling prices and inflation rates.

The royalties paid on the main production areas of the southern North Sea and South Coast were, in 1983, 14.38p per tonne. A few areas close to ports attract a higher rate. In Liverpool Bay and the Mersey Estuary the rates are lower, reflecting the very keen market on Merseyside. Royalties are payable irrespective of whether the dredged material is landed on the Continent or in the UK.

4.2.2 Dredgers and Dredging Costs

The cost of dredging for marine aggregates is extremely dependent on the cycle time. Hence small dredgers operating on the South Coast, the Bristol Channel and Liverpool Bay, where ports are very close to the dredging grounds, operate efficiently within one tidal cycle (DoE,1986). With larger vessels dredging more distant grounds this cycle time increased to 2 tides which, coupled with low fuel costs prior to the 1973 fuel cost explosion, still enabled a reasonable performance to be achieved.

In recent years however, circumstances, particularly in the southern North Sea, have adversely changed. **Fuel costs** have become of major significance and deteriorating reserves in the Thames Estuary require vessels to steam to more distant areas such as Cross Sands, increasing the cycle time to 3 tides.

The major part of the increase in the cost of landing sea-dredged materials since 1973 is due to the dramatic increase in the price of fuel oil. To illustrate this, in 1973 the price of gas oil was £13 per tonne in the UK and represented as a cost constituent on a tonne of gravel landed only 4p per tonne on a 24 hour cycle. By 1985 the gas oil price was £200 per tonne which represented a cost constituent of 76p per tonne of gravel on a 36 hour cycle.

The cost constituent on a tonne of sand or gravel will vary as a result of both the location of dredging ground and the type of material won. For example an inshore sand delivered into Liverpool or Cardiff will be relatively cheaper than an inshore sand and gravel delivered into Portsmouth or Southampton. This is because sand extraction produces less wear of dredging gear (compared with sand and gravel mixed cargoes) which keeps maintenance costs down. With more distant dredging grounds, such as those to serve London, the cost of both sand and gravel will be higher despite the use of modern, more fuel efficient vessels. In addition, wharves further upstream on a river, such as the Thames, will have higher tonne/fuel rates than those in the lower reaches because of the extra steaming distance involved. Geographical variations of the price of gas oil in the UK are not significant.

A further illustration on the effects of fuel prices can be highlighted by stating that as a cost constituent it now represents the greatest percentage part of the aggregate sale price ex-ship and exceeds both wages and maintenance costs combined, whereas in 1973 it represented no more than 3.5% of the sale price of the material ex-ship.

The rise in fuel prices is a variable cost over which the dredging industry has no control. Oil companies purchase fuel in dollars and sell in sterling. Any fall in the value of the pound against the dollar is the most significant factor affecting the cost of fuel. Inevitably, these costs have to filter through to the recipient company and eventually the general public. Rises in fuel costs are not directly related to changes in the ex-wharf price of a product since there is fierce competition from other producers, especially if return loads by lorry with other commodities can be arranged (Section 4.2.4). In an attempt to reduce vessel fuel costs, one company uses its dredgers for return cargoes of blastfurnace slag after delivering sand and gravel to ports on the Continent.

Fuel prices are no longer significantly different from those on the Continent now that the UK is in the EEC. Prior to membership of the EEC it was beneficial to take on fuel in Rotterdam etc.

Vessel costs have also increased considerably: the construction of a new 4500 tonne carrying capacity dredger built for delivery in 1973 would have cost some £2.4M. In comparison current valid tenders (excluding any special subsidies and equated on a like for like basis) would now give a figure of £9.5M, a rise of 296%. Although this percentage increase is not in excess of the RPI, another important factor to be considered is that the vessel built in 1973 was designed for a cycle time not exceeding 24 hours. With diminishing reserves attainable within this cycle time, the majority of materials now extracted come from licensed reserves that need a 36 hour cycle, undertaken in more hazardous sea conditions. While capital costs have not advanced significantly more than the RPI, the annual tonnage landed by the vessel has decreased by 32% by virtue of longer steaming times. Therefore, whilst fixed costs have risen in relation to the RPI the ratio of these costs to revenue has increased significantly. In an attempt to reduce costs per tonne because of these greater steaming distances, crewing levels are an important consideration (see later in this Section).

One of the major considerations facing companies involved with winning marine sand and gravel is deciding when to invest in a new vessel. **The majority of companies are NOT prepared to commit finance unless they have assured licensed reserves.** A few companies see investment in new dredgers as indicating good intent in anticipation of further Production Licences.

Another cost factor is **port dues** which are dependent on the port of delivery and vary from as little as 2p per tonne to as much as 80p per tonne. This can be very significant when marketing in competition with land-won aggregates.

As indicated in Section 11.2.2 pilotage costs are not normally incurred by the dredging fleet and are in any event a variable factor depending on the GRT and the port authority. Pilotage and the future role of Trinity House is currently the subject of a Government Green Paper. The implications of the latter are uncertain and this is a situation which the dredging industry will wish to monitor in case pilotage becomes either more expensive or widespread.

As an illustration Table 4.2 compares the position between 1973 and 1985 for a similar size dredger and also the comparative position between the two dates assuming that licensed reserves continue to be available within the 24 hour cycle time.

Vessel berthing

	Constructed & Operating 1973 on 24 hr cycles	Constructed & Operating 1985 on 24 hr cycles		Constructed & Operating 1985 on 36 hr cycles	
	Cost	Cost	% inc over '73	Cost	% inc over '73
Capital Cost	£2,500,000	£9,500,000	380%	£9,500,000	380%
Fixed Costs (p.a.)	£ 225,000	£ 950,000	420%	£ 950,000	420%
Depreciation (p.a.)	£ 156,000	£ 594,000	380%	£ 594,000	380%
Fuel (p.a.)	£ 50,000	£ 775,000	1550%	£ 650,000	1300%
Delivered quantity	1.25m t.p.a.	1.25m t.p.a.	-	0.85m t.p.a. (-32%)	
Average Landed Price (excl port dues)	£0.90p tonne	£2.75p tonne	306%	£2.75p tonne	306%
Revenue	£1,125,000	£3,437,000	306%	£2,337,000	208%
Gross Margin	£ 694,000	£1,118,000	-	£ 143,000	-
Return on investment (pre o'hds & finance charges)	27.8%	11.8%		1.5%	
RPI	100 (Jan 1974)	366 (March 1985)		366 (March 1985)	

TABLE 4.2   COST & EARNINGS USING A 4500 tonne CAPACITY VESSEL

Source: BACMI and SAGA.

Figure 4.5 illustrates the industry's investment in new dredgers since 1960 expressed in terms of both total carrying capacity and mean capacity for each year. It will readily be seen that peak capacity in shipbuilding occurred in 1974 with eight years (1976-79 and 1981-84) when no investment was made. Whilst expressions of interest in new dredgers are made by the Industry there is a general reluctance to place orders without the security of adequate quantities of licensed reserves to ensure the viability of new operations. Orders for two new dredgers (both with a carrying capacity of approximately 4,600t drained cargo) and each costing £7M, were placed in 1985 by the same company.

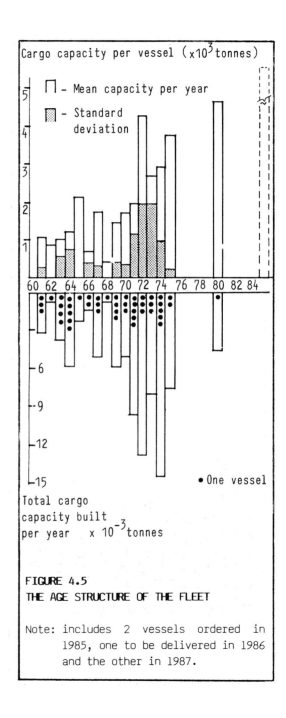

**FIGURE 4.5
THE AGE STRUCTURE OF THE FLEET**

Note: includes 2 vessels ordered in 1985, one to be delivered in 1986 and the other in 1987.

Dredging in rough weather

The cost of new dredgers is normally amortised over a period of 15/20 years. The life of a dredger will exceed the period of amortisation and depending on company policy it should be in service for 20-30 years provided a major refurbishment is carried out after a period of 20 years. Currently there are six vessels which are 30 years or more old, all working in sheltered waters (DoE, 1986).

Some of the small dredging companies, who operate in sheltered waters and who cannot afford the high capital cost of new vessels, have a policy of buying up old ships for a relatively small amount of capital outlay. These ships will be expected to operate for approximately ten years or until no longer safe or become too expensive to repair. Such vessels are cheap since they operate only in sheltered waters and therefore do not require a classification certificate of sea worthiness (ie these vessels are not fully classed).

Even with the larger companies, the cost of building a small capacity dredger (say up to 800 tonnes cargo) does not merit an expenditure of up to £2M. Preference would be given to the conversion of coasters.

Dredging companies attempt to maximise dredger utilisation to obtain a profitable return on their investment. It should be stressed here that a dredger is an expensive, **depreciating asset** which of necessity should be kept fully operational. With dead rents payable on all licensed areas (Section 4.2.1), an annual financial incentive is provided for exploitation of remaining reserves. When compared with the situation onshore, land-won sand and gravel operations are normally associated with less expensive capital investment in machinery for extraction and the transport of material to a plant site/marketing point. In addition, remaining onshore permitted reserves are sometimes seen as an **accruing asset** to be husbanded.

In a depressed market companies with interests both onshore and offshore may choose to husband their land-based permitted reserves to enable further use of the capital invested in dredgers. To that extent some quarries may reduce their outputs or temporarily close to improve overall company efficiency in aggregates production.

To achieve the best return on dredger investment the maximum use of each ship is necessary. Apart from Christmas and statutory holidays, most dredgers will attempt to be fully operational 24 hours per day on some 350 days per year.

In a reasonable year a ship will actually operate with about 20% downtime, due to annual refits, operational delays, breakdowns and poor weather. Bad weather may slow but rarely prevents steaming, but it affects harbour entry and ability to dredge (particularly anchor dredging). A ship operating on a 12 hour cycle would expect to complete about 550 deliveries, 290 trips if on a 24 hour cycle and 200 trips if on a 36 hour cycle (note only a few vessels work exclusively on a 36 hour turnround, such as the low profile 'Bow' vessels that steam to Cringle wharf in central London).

At the end of a company's financial year the total tonnage landed by each vessel is all important. Quick discharge times can be crucial both to catch a tide and save fuel and in some instances, notably the Bristol Channel, to be the first dredger on station where shared licences are involved.

The financial return on a dredger can sometimes be enhanced by extending the vessel amid ships to increase its carrying capacity. The economic importance of this is that the ratio of cargo capacity: crew increases. Several dredgers have been extended in this way. Since the number in a ship's crew reduces relatively as the size of vessel increases (DoE,1986) many companies are looking to larger vessels because the cargo capacity: crew ratio can show a higher productivity per man employed.

Although larger vessels mean fewer trips for the equivalent landed tonnage, small vessels are sometimes **necessary**, not only because some wharves have small stocking areas but also because of navigation to the wharves along shallow, narrow channels. Many of the South Coast wharves (with the notable exception of Southampton) are served by small ships (Section 12.2.3) whereas economics dictates that large vessels must be used to serve the London market along the River Thames and Newcastle along the River Tyne, for example.

Other ways of maximising dredger utilisation are through **exports** of sand and gravel to the Continent. This has been carried out for many years. Over the past twenty years the Government has considered several times the question of exports. Each time the conclusion has been that it would not be in the national interest to seek to restrict exports. The dredging Industry's capital costs are high and viability depends upon maximum and economic use of its resources. The selling prices on the Continent are normally somewhat less than in UK markets. It is therefore in the Industry's interest to land as much dredged material in the UK as possible.

In addition to serving the aggregates industry, an important part of dredger utilisation comes from works requiring bulk fill eg reclamation of marshland or harbour works (Table 4.1). A few of the smaller dredgers are occasionally employed to dredge for the calcified algae, Lithothamnion sp. off Cornwall for use as a fertilizer (S Twine, Northwood, pers. comm.). This usually involves landing only a few thousand tonnes but can help make use of a dredger especially during the winter months.

mv Bowstream
an example of a lengthened vessel

Shore plant

4.2.3 Wharf and processing costs

Marine aggregates are low priced commodities when compared with other more lucrative goods, such as containers, which provide a higher income to a port. An operator may therefore have great difficulty in securing the desired location to set up a new operation (Section 11.2).

With high levels of inflation in recent years operators can no longer look to **historic costs** for future investment in major plant. It is therefore necessary to look at **replacement costs.** This applies not only to dredgers but to processing and value-added plants.

Fixed costs will vary from wharf to wharf but could include the loan repayments for the acquisition of the site, cranes, processing plant, concrete and/or asphalt plants, infra-structure works (such as access roads, concrete aprons, storage bays/hoppers, workshops, office and weighbridge and the provision of services: water, electricity, etc). It can be anticipated that in 1985 a plant capable of producing around 200 thousand tonnes per annum would cost some £0.75M. The cost of land acquisition and wharf construction may add a further £0.5-2M to the investment. Larger processing capability will involve a considerably greater total outlay; a smaller outlay would be involved for those wharves landing only sand and selling this 'as dredged'. Processing plant is normally amortised over 12 to 15 years. However, coastal locations of wharves lead to greater corrosion wear and maintenance costs compared with land-won operations.

The Consultants have been told that the economic recession has placed several wharf operations in temporary financial difficulty due to a lack of demand for the products. With the high capital investment involved these wharves have been committed to continuing through a non-profit making period. Artz (1975) highlighted the need for a break-even point (BEP) which was defined as 'production expressed as a percentage of the total capacity of a plant necessary to reach a profitable state'. Some wharves have probably been operating below their BEP but where competition from land-won aggregates declines, or the local demand for marine aggregates increases, wharves can move back into a profitable state. The BEP will be different for each operation depending on the level of fixed and variable costs against the throughput tonnage. It can be applied to both dredgers and wharf systems. High throughputs at wharves are necessary to maintain profitability.

4.2.4 Price Structures

There is no fixed selling price for sand and gravel. Pricing is left to individual companies having regard to the market forces of demand, availability in both quantity and quality of comparable products either from other wharves or from land based pits.

In general, the cheapest dredged material will be fill for reclamation works, followed by ballast as dredged (BAD), then processed sand, with washed and graded gravel being the most expensive.

Price variations will occur within the same product, for example depending on the distance and location of dredging grounds to a wharf. Thus material dredged from shallow inshore waters, such as in the Solent area of the South Coast and delivered into Portsmouth or Southampton will be cheaper than the similar material dredged from Cross Sands and delivered into London which in turn is cheaper than material dredged off the Humber and delivered into Newcastle.

The **delivered price** of an aggregate will be the major influence on how far marine dredged material can penetrate a local market. Apart from competition from other marine producers much will depend on haulage rates in relation to the location of land-based pits and their ability to meet demand.

Ex-pit prices are usually lower than ex-wharf prices but marine landings often have the locational advantage in that the wharves are within the main areas of demand whereas pits will normally be several kilometres outside urban areas. Thus haulage costs from wharves will generally be less than those from pits. This point is illustrated in Figure 4.6.

Haulage rates tend to be **higher** for deliveries over short distances from a pit because the majority of sales will be into the main urban areas. Conversely, rates from wharves will be **lower** for the first few kilometres. Using the example in Figure 4.6, the point at which the haulage price would be the same (@ £1.75 per tonne) for a pit and a London wharf 32km apart is 18km from the pit and 14km from the wharf.

Penetration of the market area by marine and land-won sand and gravel will be influenced by the respective production rates, haulage costs and possible user prejudice against either source (Section 7.2.2). The determination of haulage rates will be influenced by fuel costs, the size of vehicle used and the road conditions in the vicinity of the pit or wharf. Some producers offer a flat rate irrespective of distance. Others charge on actual km travelled by a lorry or on cost bands of radial km (eg at every 5km) to a customer. Where different road conditions are found near a

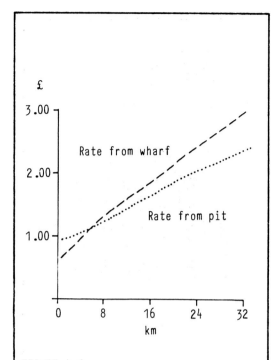

**FIGURE 4.6
GRAPH SHOWING TYPICAL HAULAGE RATES FROM A PIT AND A WHARF**
(BASED ON LONDON)

Source: BACMI

Lorry transport

pit or wharf, an operator may reflect this by charging different rates in particular directions.

Special grades of material eg fine sand can be transported in excess of 160km to meet a specific need but generally sales will be confined to within 50km of a wharf and less in totally urban situations. Virtually all aggregates whether rail-borne, sea-borne or transported by waterways will end their journeys by road (with a few exceptions such as pump discharge for marshland reclamation).

Over the past 10-15 years there has been a move towards transporting aggregates in larger and fewer lorries. This has resulted in an increase in eight-wheeled tippers having a 20 tonne payload, and a decrease in the number of four-wheeled tippers. The 'Artic' has now virtually disappeared. Singleton (1981) saw these trends in vehicle size changing due to comparative vehicle costings which showed that the cost per tonne of four-wheelers is 40% greater than eight-wheeled vehicles.

A few mineral operators undertake **back loading** (ie sending out dredged sand and gravel and returning with stone, grain etc) to keep transport costs down. Although the industry do not consider this involving a very significant proportion of sales it is practised by both large and small companies. The smaller companies in particular offer very keen haulage prices which affect other operations in an area. Although back loading seems likely to continue its widespread use is very unlikely. Some established land-based producers proposing to enter the marine sand and gravel market for sales are considering back loading as possibly the only viable means of extended market penetration (Sections 4.4.1 and 11.2.1).

Rail transportation costs are negotiated by an operator with British Rail and are incorporated into a Freight Agreement. These rates take into account the anticipated tonnage and an operator's commitment to long term movement by rail. Lower rates are given where high tonnages over a long period are proposed. The rates incorporate overheads, including a proportion for track and bridge maintenance (cf. a road fund licence) even though there may be several other users of a line at anyone time. In the longer term some users may disappear (C R B Goldson, BR, pers. comm.).

British Rail also take into account equivalent journeys by road to the area of the receiving depots since this will involve lorries in negotiating different driving conditions, as indicated above. Freight rates, which are index linked, are reviewed at six monthly intervals. Financial penalties are imposed by British Rail if an operator does not move a minimum guaranteed annual tonnage (cf. CEC dead rent on production licences referred to in Section 4.2.1). This minimum level will be set as high as possible by British Rail, whilst an operator seeks an agreed lower level.

Rail haulage of aggregates had previously only been encouraged by British Rail for distances in excess of 100km but in a more competitive economy shorter rail movements have become commercially attractive, such as Newhaven to Crawley, a distance of 40 km which can compete with road haulage. The transport of marine sand and gravel by rail represents about 7% of total UK landed material (Section 11.2.3).

In addition to the list of component costs in Table 4.3, out of profits (mainly from offshore operations) will come the repayment of **development costs** which will include prospecting (DoE,1986), seeking planning permission, CEC licences and DTp consent. The lead time necessary for any return on these costs can be several years and such lead time costs money.

4.3 GRANT AID

4.3.1 Ships

There is no direct grant aid to the industry for dredgers. The scheme used, known as the Shipbuilding Intervention Fund, is administered by the Department of Trade and Industry and conforms to the EEC Directive 'Aid to Shipbuilding'. The scheme is limited to vessels of 150 GRT or more and fully classed. It is not operated through the owners of ships but by the shipbuilders who apply for financial assistance. The object of the Intervention Fund is to enable UK builders to break-even on construction costs where there is non EEC competition.

It is important to note that the figures given in Section 4.2.2 on vessel costs are net prices ie after making allowances for monies from the fund which will vary from time to time. The Fifth Directive of this scheme was effective from 28 April 1981 and lasts until 31 December 1986. Shipbuilders under the scheme currently receive up to 20.5% of the contract price and a further 2%, known as shipbuilders' relief, for payments of VAT etc. Thus a £9.5M dredger to an owner would actually cost £11.6M overall, with the balance coming from the Department of Trade and Industry.

The Sixth EEC Directive for the period 1987 onwards is likely to be more restrictive (J A King, DTI, pers.comm.).

4.3.2 Waterways

At present the marine sand and gravel industry have not made use of grant aid under Section 36 of the Transport Act 1981. This refers to the Inland Waterways Freight Facilities Grant Scheme which provides aid where the facilities relate to the carriage

---

OFFSHORE -
    CEC royalties
    fuel
    wages
    maintenance
    port dues
    discharging
    insurance
    profits
    (loan repayments/capital
    amortisation for dredger)
    (pilotage)

ONSHORE -
    local authority rates
    water rates
    electricity consumption
    wages
    maintenance
    profits
    insurance
    transport costs
    (loan repayments/capital
    amortisation for plant, buidings
    etc)
    (processing)

TABLE 4.3
COMPONENT COSTS OF DELIVERED MARINE SAND & GRAVEL

(words in brackets do not apply to all operations)

of freight by inland waterway, including cargo carrying craft. Although there is no precise definition of "inland waterway", the Secretary of State does not propose to offer grant aid downstream of any point where the opposite banks are more than 5km (at high water spring tide) or 3km (at low water spring tide) apart. Facilities for traffic in sea-going vessels which are to be provided upstream of these limits will only be considered when the traffic moves a substantial distance within the inland waterway. All of the existing wharves on London River fall within these limits.

This grant does not therefore extend to dredgers and existing wharf facilities but does cover, inter alia, barges, wharves, handling equipment and storage areas for transhipped material. Eligibility for grants are given where fixed facilities (eg wharves) are to be operated for a minimum period of 5 years, and 10 years for cargo carrying craft (eg barges). The amount of grant available, in relation to the specific project, will be influenced by the amount of environmental benefits a scheme can produce, primarily related to removing lorry traffic.

As regards possible extension of Section 36 grants to cover aggregate dredgers and existing wharves, the Consultants see this as not only assisting Central Government with their policies to encourage landings, but also offering the industry a financial incentive to improve the environmental appearance/operation (and possible relocation) of some wharves.

The amount of grant aid would be dependant upon the extent of environmental improvement likely and similar procedures to those operating at present for rail and waterway grants should apply, including consultation with local authorities likely to benefit from any improvements.

As regards to vessels, **the Consultants believe that it would not be in the interests of Central Government to consider grant aid for dredgers, per se, because of the fiscal implications of extending such grant aid to merchant vessels generally.**

There would appear to be potential for barge transhipment, although investment costs seem likely to be high even with possible grant aid assistance. Whilst a new fleet of barges might be precluded on the grounds of costs, there would seem to be scope in exploring with established barging companies some form of joint enterprise either by direct contract or by partial investment in new barges with motive power and labour provided by the contractors. The use of barges for marine aggregates is considered further in Section 11.2.2.

4.3.3 Rail links

Section 8, Railways Act 1974 as amended by Section 16, Transport Act 1978 relates to the Rail Freight Facilities Grant Scheme and provides aid for the carriage of freight by rail, or the loading or unloading of freight carried by rail. This includes the construction of new rail handling equipment, locomotives and wagons. The grant can be applied to wharves, whether or not an existing track is located in the immediate area. The existing wharves at Cliffe and Newhaven already had sidings to their sites and were also established before Section 8 Grants were available. Grant aid for new receiving depots has been given under Section 8 and for rolling stock under Section 16 in connection with the movement of marine aggregates. Since 1985 all rolling stock has to be privately owned.

Those companies with wharves which are rail-linked will probably continue to make use of grant aid for rolling stock and new distribution depots. There seems generally limited opportunity for the marine sand and gravel industry to make wide use of available grant aid for many new rail-links.

Additional wharves likely to move marine sand and gravel by rail (and the application of grants) seems limited to South East England where rail-links could connect with outlets (mainly concrete batching plants) in the London market place. One proposed wharf and an existing wharf on Thames-side are rail-linked and could be utilised if both demand and supply of marine aggregates were assured in the longer term.

4.3.4 Other aid

Other Government backed grants, such as the Regional Development Grant (RDG) are not thought appropriate to establishing wharves for the landing and processing of marine sand and gravel. The RDG applies to Assisted Areas only and relates primarily to job creation although some selective assistance is also available for projects which protect existing employment but which would otherwise not go ahead. With low manning levels required to run a wharf there seems little prospect of that the RDG will be a financial incentive to establish new wharves in such areas.

4.4 MARKET INFLUENCES

4.4.1 Intra-Company Policies

In general the financial rate of return is lower in the marine sector than with land-won aggregates. None of the major companies relies solely on income from marine

Loaded train leaving Cliffe

aggregates. Therefore the price of land-won aggregates will continue to be a controlling influence on marine sales. Pricing is not influenced by any encouragement for landings given in the National Guidelines (Circular 21/82, Welsh Office Circular 30/82). The problems of high costs associated with wharf operations are identified in Section 4.2.3.

Companies with interests in both marine and land-won sand and gravel will be influenced by their ability to secure planning permissions for new or extended pits. Some companies have an aggressive marketing policy ie to sell their products as fast as they can be dug out of the ground but the majority now see the wisdom in cutting back production (brought on originally by the economic recession) so as to conserve supplies in the face of the difficulties in obtaining further planning permissions and to allow undug reserves to accrue in value (Section 4.2.2).

The marine industry has thus been able to take advantage of this situation as their share of the UK sand and gravel market has been sustained in recent years at between 12-13% (DoE,1986).

While dredging companies who are associated with receiving/marketing companies will be looking to make separate profits much will depend on the intra-company transfer prices. It appears to the Contractors that there is a degree of flexibility over where a large organisation can place its greatest profit margin. In the marine sand and gravel industry profit margins are most vulnerable at the receiving/marketing end, as in concrete, because this is governed by the ex-ship price of the aggregate. Market influences of demand and effects of local competition will also affect profit margins.

4.4.2   Inter-Company considerations

The penetration by a new operator in an area in which marine aggregates are already landed may be difficult. This applies not only to companies currently involved with receiving landed material elsewhere but also to land-based producers who have considered expansion into the processing and marketing of marine aggregates, but an aggressive approach can succeed.

As with trade and industry generally, the operation of free market forces allows prices to be adjusted to suit local conditions. This situation also applies to the sand and gravel industry where some views have been expressed that prices can sometimes be adjusted to reflect competition, whether from land-based producers or within the marine industry. The effect of this could be that some marine operations close, others fail to become established, and some succeed in penetrating the market area.

IMPROVEMENTS IN DREDGING TECHNOLOGY

> **5**
>
> CONTENTS:
>
> **Ship design**
>   Seaworthiness
>   Economy of operation
>   Flexibility
>
> **Dredging and dredge pump development**
>   Dredging in deep water
>   Dredging beneath overburden
>   Dredge-pipe articulation
>
> **Production monitoring techniques**
>   Cargo loading rates
>   Cargo quality

A description of the aggregate dredging fleet and its operation in 1985 is given in DoE (1986).

5.1 SHIP DESIGN

5.1.1 Seaworthiness

In terms of the development of **hull design** and **sea-handling characteristics** aggregate dredging vessels are undifferentiable from the mercantile fleet generally. No revolutionary design alterations are heralded for the forseeable future, and new aggregate vessels will continue in the tradition of minor improvements and 'polishing' of design.

A possibility for design improvement has been identified in relation to **deck gear.** Two trends in aggregate dredger development, those of installation of self-discharge gear and of operating long hauls between dredge and landing sites, have conflicted to some extent. Thus whether a ship can steam into storm seas is often dictated by the threat of damage to vulnerable deck gear. There would appear to be considerable scope for the design of streamlined, strengthened, more water resistant and simplified self-discharge equipment.

5.1.2 Economy of operation

Whereas in many instances the development of aggregate dredging technology can be seen to follow the innovations of the engineering dredging fleet (see below), a major difference occurs in terms of speed versus economy of operation. Although turnround time is important in aggregate dredging cycles, it cannot be so at the expense of economy*. Thus in the Consultants' opinion hull resistance and engining criteria will continue, unless there is a radical change in the economics of marine aggregate production, to develop along lines of optimum economy. Research, such as that conducted at IHC in Holland (Figure 5.1) will continue to refine propulsion costs.

Crew costs are another important area where economies have been effected in the past (Section 4.2.2) and will continue.

> *Unlike construction industry dredging, where rapid completion of projects during fine weather windows is normally or paramount importance. Thus these vessels normally have larger engines and pumping capacities than aggregate dredgers.

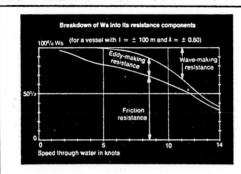

FIGURE 5.1
ANALYSIS OF HULL FRICTION COMPONENTS

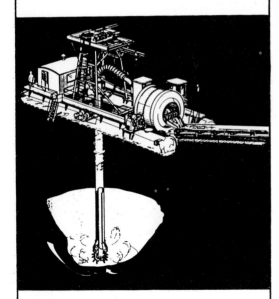

FIGURE 5.2
AIRLIFT MINING DEVICE

5.1.3 Flexibility

Although in terms of operational efficiency aggregate dredgers cannot normally compete with dredgers specifically designed for engineering projects, the distinct advantage of access to local aggregate/fill reserves often outweighs other cost considerations, and one-off projects such as reclamation, beach nourishment and coastal or offshore substrate stabilisation play an important role in present day viability of aggregate dredging operations. If this situation persists, or the importance of non-wharf deliveries increases still further, flexibility in terms of ship manoeuverability and discharge capabilities may be of distinct advantage. The possible increased development of return cargoes in the future, with associated handling requirements, also enters into this consideration.

Increased flexibility means increased ship construction costs. For example, for a ship to have a pump discharge facility may put 20% on the cost of the basic hopper dredger. Whereas a pronounced change in marketing patterns would be required for investment in features such as twin-screws, or multiple anchor winch systems, it is envisaged that pump discharge or bottom-opening valves could well repay investment; a lot of money is currently being spent in Great Britain on feasibility studies relating to projects such as offshore metalliferous placer mining, barrage construction, offshore energy generation, major reclamation and coastal protection schemes and of course the Channel Tunnel (Section 12.2.4).

Decreased ship flexibility is also a possibility in the form of offshore pumping units served by hopper barges. The concept of a **mothership** constantly on station dredging for sand and gravel with sea-going barges plying to and fro to the shore has attracted the attention of both mineral operators and planning authorities for many years. Such a system could make full use of dredger utilisation and economic models have been devised to demonstrate its merits (Artz,1975).

There are a number of technical problems to overcome in the implementation of such a system, not least of which is ship to barge cargo transfer during bad weather, which would require the development of dynamic positioning capabilities.

It is felt within the Industry that this type of system would only be given further active consideration if there were a substantial increase in the 13-14% contribution that marine material currently makes to the total UK demand for sand and gravel.

On a somewhat smaller scale it can be envisaged that with the demise of many of the older small dredgers, some of the smaller wharves would have to close (and maybe lose valuable markets). A possible alternative is that they continue to be served using larger dredgers (both existing and proposed) which could steam into sheltered waters, such as The Solent or Bristol Channel, and discharge their cargoes into barges (see also Sections 12.2.3 & 12.6.2).

A similar type of operation was used in beach nourishment at Hayling Island, Hampshire in 1985 where sand and gravel was dredged from Owers Bank off Worthing and brought to about 1n mile from the shore. The cargo was then pumped into barges and manoeuvred to the toe of the beach for bottom discharge.

5.2 DREDGING AND DREDGE PUMP DEVELOPMENT

5.2.1 Dredging in deep water

The use of a **suction pump** limits dredging depth to about 30m below the level of the pump. Thus even in modern large vessels where the pump is located well below the water level, dredging cannot be currently entertained on the continental shelf beyond the 35m isobath. It is apparent that a considerable proportion of existing licensed reserves lies beyond the reach of the vessels currently working them.

Gravel can be dredged at greater depths using one of three techniques:

Jet pumps' operation has been described elsewhere (DoE,1986). As a result of the poorer power input:production ratio, combined with high wear problems in the venturi pipes, there is currently a lack of interest in this type of pump.

Air lift pumps (Figure 5.2) have been extensively developed for deep aggregate working in reservoirs (Anon,1977). This technique utilises the simple principle of compressing air down the outer sleeve of a twin dredge pipe and allowing it to rise up the central pipe; in so doing suction is created and aggregate is also lifted. It is believed that this method has been tried in an estuarine environment, but did not meet with much success due to vessel motion problems. Further evaluation is necessary if this technique is to be considered for offshore production.

In-pipe pumps. Here the suction pump is mounted **within** the dredge pipe, instead of in the ship, (Figure 5.3). This technique has been developed for deep excavation and improved solids delivery by the engineering dredging industry, and will be used on the new dredgers currently under construction for ARC Marine (giving the latter vessels a dredging capability of 45-50m). Problems associated with this development are outboard

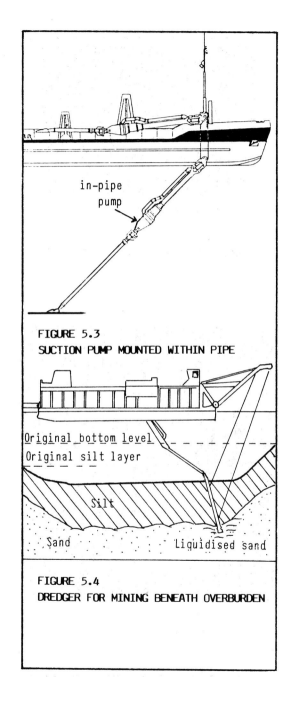

FIGURE 5.3
SUCTION PUMP MOUNTED WITHIN PIPE

FIGURE 5.4
DREDGER FOR MINING BENEATH OVERBURDEN

weight, sealing problems with submerged equipment, and the difficulty of repairing the pump or clearing blockages. If, as with the new aggregate dredgers, only a single in-pipe pump is used, pump discharge is precluded (some engineering dredgers overcome the problem by having two pumps operating in tandem).

5.2.2 Dredging beneath overburden

Winning of marine aggregate from beneath overburden (usually fine sand or mud) is not at present economically viable in Britain. High grade deposits are however known to exist in such situations.

The impracticability of working beneath overburden relates to the cost of dredging the covering layer and dumping it elsewhere. A method for mining sands **from beneath** a fine sediment overburden has been developed in Japan (Figure 5.4). A high pressure water-jet system fluidises the underlying sand, and a low pressure diluting water system controls the fluidisation and lift of the material to the surface. The mud, which is of lower specific gravity than the sand, remains 'floating' in situ on the surface of the fluidised sub-layer. Sand uncontaminated by overlying mud can therefore be extracted. It is not clear whether such a system could cope with gravel material, or open-sea conditions. It is not thought that pressures on reserves could lead to consideration of this form of dredging in Britain within the next 15-20 years.

5.2.3 Dredge-pipe articulation

Presently used trailer dredge-pipe articulation and control systems (DoE,1986) rely to a large extent upon vessels steaming parallel to the tide for efficient and damage-free operation. This inability to readily cope with 'crabbing' across the seabed requires that mining patterns are controlled by hydrographic constraints rather than the geology of a deposit. Future dredging management may require a change in this situation (Section 6.3.1), which will necessitate technological development of pipe articulation hardware and control systems.

5.3   PRODUCTION MONITORING TECHNIQUES

5.3.1 Cargo loading rates

In this field the aggregate dredging industry is following the lead of engineering

Typical modern dredge-pipe articulation system

dredging technology, and on the newest vessels instrumentation and automation systems have already been adopted (DoE,1986).

In the coming years it is to be expected that computer controlled and monitored systems will replace the electro-mechanical devices which currently prevail.

It should be stressed that there is currently a great divergence between the level of control operated on the newest and older vessels, and there is considerable scope for updating the systems in use on most vessels.

5.3.2   Cargo quality

Cargo quality is clearly of primary importance in aggregate dredging, whereas in dredging for navigation or construction purposes the cargo is invariably dumped as waste*. Observation of the colour of the incoming cargo, and spot sampling, are currently relied upon for quality control (DoE,1986). Presumably at least partly because of the lack of stimulus from the engineering dredging industry, no instrumentation has been developed for measuring in-pipe material quality. The possibilities of such devices have apparently been looked into in the past, but abandoned as 'not leading to practical results' (Industry comment).

Technology is currently being developed for the in situ measurement of particle-size and density of sediments in the natural environment in response to coastal engineering requirements, which once established in a reliable and economic package could be utilised by the dredging industry. In the Consultants' view two important benefits would accrue from the use of such instruments:

(i)   If a draghead sensor was automatically coupled to a voiding valve immediately down-flow from the pump, cargo contamination from clay or chalk could be avoided, and

(ii) in conjunction with position fixing and data logging systems (Section 6.2.4), a continuous record of cargo grade and impurity concentration would constantly add to prospected knowledge of dredging grounds, permitting cargoes of specific quality to be more readily obtained in response to the day-to-day requirements of the wharves.

Unfortunately, the development of this type of equipment for use in the natural environment is in its infancy, and the rate that it will progress is entirely dependent

*In some restricted localities navigational dredging cargoes are landed as aggregate (DoE, 1986).

Electro-mechanical dredging controls

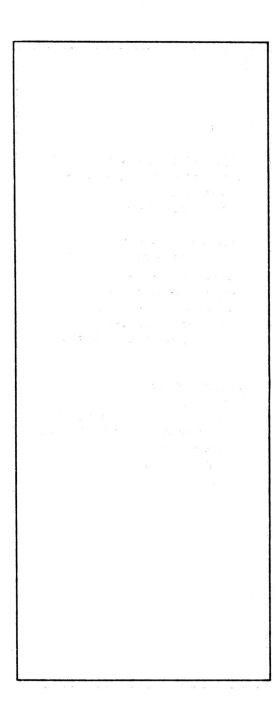

upon research funding priorities. The compatibility of the requirements of the aggregate dredging industry and coastal engineers (ICE,1985) should be established, and if coincident, joint pressure could be usefully brought to bear.

IN SUMMARY, with the exception of the adoption of in-pipe pumps (confirmed by the 1985 orders of ARC's vessels) possible use of small-scale offshore cargo transfer systems, and potential development in dredge-pipe control and production monitoring towards the end of the century (the theme of which is taken further in the next chapter), little radical change in ship and dredge technology is envisaged for the next 15-20 years. However, considering the extensive variation within the fleet to which current technology has been utilised (DoE,1986), much effort will undoubtedly be spent during this period on updating existing vessels.

FUTURE TRENDS IN RESERVE EVALUATION AND MINING MANAGEMENT

6

CONTENTS:

The potential for change
  Attitudes within industry and
    government.
  Technological evolution

Prospecting efficiency
  Current prospecting methods
  Pre-survey assessment
  Reconnaissance surveys
  Mapping and quantifying
    aggregate deposits.
  Environmental data collection
  Cost

Mining management
  Mining strategy
  Prevention of sterilisation of
    reserves by screened out sands
  Progress monitoring
  Dredging control

6.1 THE POTENTIAL FOR CHANGE

6.1.1 Attitudes within industry and government

In the Consultants' opinion, the position of the marine aggregate industry in Britain today is one which involves a degree of justification. Over the past twenty years the Industry has shown that offshore aggregate production is viable. To the observer on the sidelines it often appears (if incorrectly) that this success has been achieved serendiptiously. For the future it would appear to the Consultants that the Industry must be prepared to justify their claimed capabilities relating to areas such as;

(i) the extent to which the aggregate content of licence areas can be reasonably accurately prospected,
(ii) whether a high percentage of the aggregate identified in a reserve can be recovered, and
(iii) leaving of worked-out grounds in a reasonable condition comparative to initial environmental quality,

thus providing confidence for future investment of capital, livelihood and environmental heritage.

That Government is seeking this reassurance is recognised within the dredging industry, and the Consultants have noted that attitudes are changing accordingly. Indeed, management has to some extent been conditioned by ten years of recession in construction activity, and will have reached its own conclusions regarding future cost-effective utilisation of the nation's reserves. There is an increasing level of awareness and interest, particularly within the progressive companies, about what is happening on the seabed; a move toward utilisation of environmental technology and toward interrelation rather than competition with other users of the sea.

6.1.2 Technological evolution

Coincident with any increase in awarenes at the 'seabed' end of dredging operations, and in many ways probably a contributing factor, have been the enormous advances in **microchip technology** of the past decade. Spin-off from the space industry and development of offshore oil and gas fields have provided off-the-shelf data acquisition, storage and processing systems that have revolutionised working at sea. The innovation rate for electronic equipment is very high, and as the dredging industry becomes aware of and utilises suitable developments consequent improvement in prospecting and mining management can be expected.

In addition to utilisation of technological developments from other fields, definite **areas of need can be identified within the dredging sphere which warrant direct research by the Industry or on the Industry's behalf.** The most important of these requirements is improved ability to rapidly map and characterise sub-surface sediments (DoE, 1986), specifically coring and sub-bottom geophysical profiling. These problems are shared with the offshore construction industry, but are not so pressing in the latter due either to less extensive sampling requirements (as in coastal engineering) or the availability of large budgets (as in offshore hydrocarbons development). The problem is also shared with the worldwide placer mining industry, within which however the United Kingdom marine aggregate industry plays a leading role in coarse deposit exploration (Cruickshank, 1982). Research breakthrough in this field could thus be reasonably expected to come from the sand and gravel industry.

6.2 PROSPECTING EFFICIENCY

6.2.1 Current prospecting methods

An investigation of past and present prospecting techniques used in the offshore aggregates industry has been undertaken within this study. A factual description is presented elsewhere (DoE, 1986) covering pre-survey considerations, survey vessel and navigation requirements, seabed mapping and profiling, and deposit coring and trial dredging. An important conclusion the Consultants drew from this investigation is that there is presently considerable variation in the sampling methods, densities and expertise deployed at different localities and between different companies.

6.2.2 Pre-survey assessment

All data on Recent and Quaternary sediments of the continental shelf could usefully be collated in relation to sand and gravel resources. The exercise would ideally take the form of a publication, backed-up by a computer data bank accessible to subscribers from within the Industry. The need for continuity and updating necessitates the housing of such a service within a national organisation. BGS have already initiated a data bank of this type, and would seem a suitable candidate. It is understood that the CEC are already taking the first steps towards promoting such a service.

The collation programme would undoubtedly highlight areas of data shortage at the 'regional reconnaissance' level, which could also be usefully infilled using BGS sampling and data analysis services. This fillip to national prospecting efficiency and forward planning capabilites might usefully be funded by central government.

6.2.3 Reconnaissance surveys

The primary requirements for wide-ranging initial phases of prospecting, those of rapidly covering the ground whilst acquiring diagnostic data of interpretable quality, are currently met for bed-surface mapping (sounding, sonar, grabbing). **Subsurface profiling however has been more disappointing, as high frequency acoustic sources do not penetrate gravels well and low frequency sources, less sensitive to the acoustic impermeability of gravels, do not provide good resolution.** (DoE, 1986).

This problem has long been recognised; Kelland (1971) suggested that for aggregate prospecting a dual frequency acoustic profiling system should be developed, with two sound sources selected to optimise the opposing requirements of penetration and resolution.

For the short-term best results from presently available profiling systems can be obtained by recording raw data at sea, and processing and printing the records at a later date under laboratory conditions. In this way the effects of operator judgement can be eliminated, records showing different features under different processing control settlings can be superimposed, and signal to noise ratios can be optimised at all times, thus giving the greatest possible return from the data.

For the future there is a wide variety of research leads which appear promising. In the same way that it has been recognised that a 'multifaceted' approach to prospecting provides the optimum chance of success (DoE, 1986), it is likely that a combination of

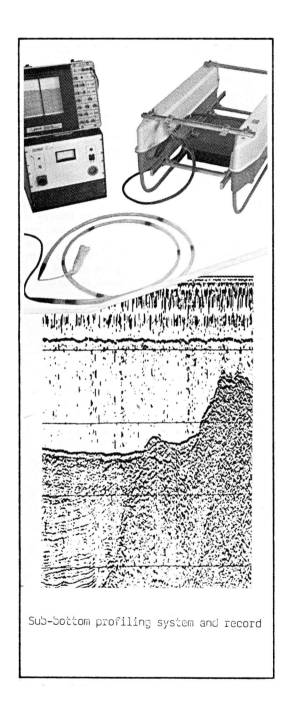

Sub-bottom profiling system and record

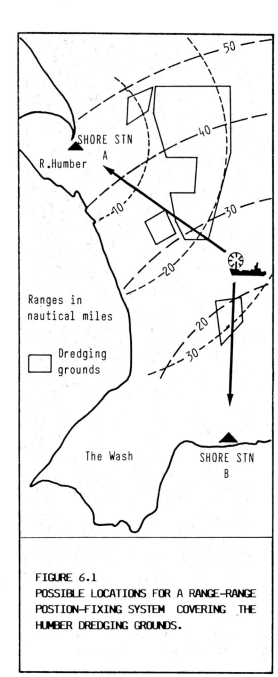

FIGURE 6.1
POSSIBLE LOCATIONS FOR A RANGE-RANGE POSTION-FIXING SYSTEM COVERING THE HUMBER DREDGING GROUNDS.

geophysical (and possibly geochemical) measurements made simultaneously will offer the best prospect for deposit identification in profile. In a brief review of recent research, the following ideas have emerged relevant to placer prospecting:

(i) Dual source acoustic systems (Kelland, 1971).
(ii) Electrical resistivity. In the United States Nebrija et al (1978) have pioneered a technique involving simultaneous measurement of acoustic profiles and electrical resistivity (with a towed Schlumberger array). They concluded that potential subsurface deposits (especially thick gravel layers), undetectable from superficial samples, can be inferred from the combined data. Research on related aspects of geophysical probing has been undertaken in the United Kingdom (Bennel et al, 1982).

(iii) Nuclear techniques. The measurement of natural radiation in the seabed has shown possibilities as a tool in the search for marine mineral deposits (Noakes and Harding, 1982). A Harwell-developed sensing 'eel', towed along the seabed, has been shown to react to the differing radioactive properties of the upper 1-2m of sediment. For example the uranium/thorium ratio showed a progressive decrease from solid rock, to mud, to gravel and to sand, due to the more abrasive-resistant thorium-bearing minerals being concentrated in the sands (Miller et al, 1977).

It is evident that there are techniques worthy of further research and development which could be of inordinate value in locating the sand and gravel reserves of the future.

Improved coring facilities would also be of use in relation to reconnaissance prospecting, and are now discussed within the context of the second phase of prospecting surveys.

6.2.4   Mapping and quantifying aggregate deposits

Once a broad area of sand and gravel deposit has been identified, prospecting effort is concentrated on **quantifying the reserve.** The trend for the future here is towards more accurate definition of the reserves, based essentially on the generation of a larger volume of data.

The preliminary requirement for any increase in accuracy is better **position fixing.** In certain situations, where a fixed structure or buoy lies close to a dredging area, parallel indexing using the ship's radar can provide accurate postioning. More flexible are a variety of well-tried shore-based micro-wave systems are available at reasonable cost, (range-range or range-bearing systems, accuracy $\pm$5m, range 50n miles - Figure 6.1). Satellite navigation probably holds the key to future position fixing. These Global Positioning Systems (GPS) are currently available, and sufficient satellites to provide 24 hour operation should be in orbit by late 1988 (Ball,1985). These compact, portable systems will give absolute vessel position to better than 15m on a 1 second rate of updating, and tidal level to better than 0.1m. The present cost of such a system is in the order of £30,000.

The choice of positioning systems for prospecting cannot be viewed in isolation, and must be related to operational mining requirements and licensing control measures (see below).

The second major requirement for more accurate deposit mapping is **data handling systems,** as large quantities of analogue records annotated manually become too unwieldy. Many micro-computer based systems are available which will log inputs from positioning and geophysical devices; these can be interfaced with computer mapping systems which will plot survey lines and data values.

A specialised version of these data handling and processing systems is currently available; **Seafloor Mapping Systems,** where side-scan sonar data are recorded, 'played back' in enhanced form and corrected for the geometric distortion inherent in the sampling system. Using this 'mosaicing' method a full picture of the acoustic properties of the seabed can be accurately displayed (Figure 6.2).

Use of accurate positioning, data recording and 'play-back' methods is more expensive than the basic systems in use today, (costed in DoE, 1986). Expense will undoubtedly decrease as time passes, following the trend of all microprocessor developments. The advantages to the operators can be summarised as increased accuracy of mapping, ability to handle larger amounts of data and enhanced data retrieval and quality - the ultimate benefit being a much more complete assessment of the licensed reserve than is currently put together.

The final, and probably most important element in deposit delineation, is **information on aggregate thickness and quality, derived from coring.** Present day reserve evaluation is typically based upon core samples taken at 1000m intervals. The efficiency with which the deposit can be defined is illustrated in Figure 6.3. Here, a series of shell and auger boreholes at 100m spacing are available for a lag gravel and

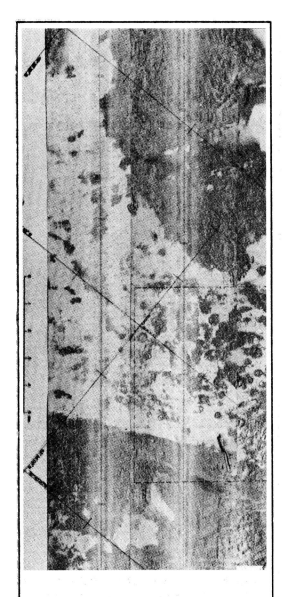

FIGURE 6.2
SMS SIDE-SCAN MOSAIC, SHOWING SAND COVER AND HARD-GROUND AREAS (dark).

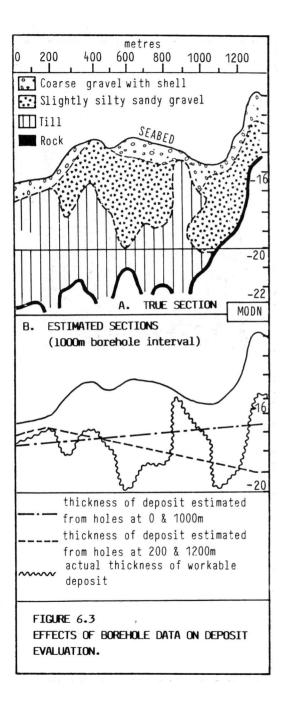

FIGURE 6.3
EFFECTS OF BOREHOLE DATA ON DEPOSIT EVALUATION.

fluvioglacial gravel deposit overlying till in the Bristol Channel. The base of the workable aggregate deposit undulates considerably although not abnormally. The effect on accuracy of taking samples at 1000m intervals instead of 100m is shown in Part B of the figure; it can be seen that the estimate of the reserve varies by about 30% according to the (chance) siting of the two selected cores. In this instance one of the approximations relates quite well to the actual reserve volume, but instances can be envisaged where gross errors can be incurred. The operational implications for differences between actual and estimated aggregate/clay boundaries are discussed further below.

Improved quality geophysical data is one way to better this situation, but the ideal solution is undoubtedly a far greater density of boreholes. Two points can be made relative to this requirement:

(i) Even using existing vibrocoring methods, with costs estimated at £200 per metre of core (DoE, 1986), the expense of adequately sampling proposed dredge areas is not particularly excessive. For example, consider the annual tonnage of material dredged in the U.K. If, allowing for rejected material, 25M tonnes is raised, this represents about 14M m³ per year. Assuming an average deposit thickness of 2m, this equals an annual mined area of 7M m². If this area had been prospected with a core every 200m, some 175 holes would have been drilled, at a cost of about £70,000. Related to the aggregate won (15M tonnes) the cost is about 0.5p per tonne.

(ii) The idea of increased coring density would obviously be far more attractive to the Industry if this form of sampling became more cost-effective. The need for new technology is recognised by the offshore construction industry and academic circles, but no new ideas have recently been put forward. An ideal 'corer' would be lightweight and simple (ie with low mobilisation costs) capable of sampling gravel to at least 3m depth in a time of about 15 minutes.

In relation to the latter, the unit would ideally be autonomous, thus being placed untethered on the seabed prior to coring and retrieved afterwards, obviating the time consuming need to anchor. There is scope for original thought and research here, particularly considering the economic return which could be forthcoming from a practical invention. Lines of development which have been suggested include augering devices, air-lift systems as used by divers (Wilkes, 1971), and super-freezing cores around narrow coolant conduits driven into the seabed, as are currently used on a minor scale for coring gravels in rivers (Walkotten, 1976; Carling and Reader, 1981).

6.2.5 Environmental data collection

The possible advantages of dredging companies generating 'environmental' data is covered in Chapters 9 and 10, on coastline stability and fisheries respectively, where the need for such information is discussed at length. It is pertinent to raise the subject within the current chapter for two reasons:

(i) As stressed subsequently, this type of information (eg current meter data, turbidity records, underwater photography of the seabed, biological surveys) would normally be most economically collected at the time of prospecting. For example recording current and turbidity meters can be deployed at one or two sites in the final prospecting area over a week whilst reserve mapping is carried out; underwater photographs can be taken simultaneously with grabbing with no loss of time and grab samples can be quickly inspected and sieved at sea for faunal content if a biologist is aboard (White et al, 1974) and the sample retained intact for aggregate-related laboratory analyses. As will be made clear in Chapters 9 and 10, it is the Consultants' view that generation of this type of data would improve relationships with other users of the sea, and will help prevent frustrating delays during licence applications which occur when there is insufficient evidence to support or refute objections to gravel extraction.

(ii) Environmental data may be directly relevant to reserve management. Examples that can be cited are; underwater photography would reveal the extent of unsuitable thin surface layers that may have to be removed prior to production dredging (shell or other biological material), or the presence of large cobbles which would impair trailer dredging efficiency; biological assessment would reveal the presence of sandeels (Section 10.3.2) which would necessitate careful timing of initial dredging; current direction and speed data would be necessary if any attempt was to be made to prevent over-sanding (see below).

Retrieval of current meter

6.2.6 Cost

As with any call for increased precision of measurement so that industrial operations may be more efficiently and sympathetically run, a major objection arises in the costs involved. Whereas the Consultants suggest that such cost increases can be offset to a considerable extent by improved reserve recovery, operational efficiency and success in licence application, two general points need to be made:

Anchor dredge pipe

(i) With environmental data collection, the scale of any programme needs to be closely tied to the expected level of return from the project.

(ii) More detailed prospecting data generation is needed for investigation of licence applications as well as for reserve management. However, it is a costly input in the instances where a licence is refused. The CEC might encourage more detailed prospecting by offering partial recompence of prospecting costs for failed licence applications.

6.3   MINING MANAGEMENT

6.3.1   Mining strategy

The present methods of dredging have been developed over a period of 30 years by companies who are foremost in the dredging field (DoE, 1986). In the early days science and technology did not play a large role in the dredging industry, and a far greater measure of reliance was placed on the ability of the Master than is commonplace in the modern trailer dredgers of today, which are becoming equipped with the latest devices for controlling the operation. Whereas trailer dredging is, wherever possible, favoured, there is still a requirement for stationary dredging in a number of areas where trailer dredging would be impractical because of the configuration of the deposits.

The activities of the Master of a dredger are today far more closely controlled by companies than they were in the past, and this trend is continuing. The decision on where the material is going to be dredged is the joint responsibility of Management and Master. The Master pays close attention to the material being dredged to ensure that it satisfies the customer's requirement. He relies on initial prospecting data (of widely varying quality), and his own experience, to define production areas. It is the Consultants' observation that, in practice, this results in each Master having a series of spot locations (for anchor dredging) or courses (for trailer dredging) where he knows he can find a specific type of good quality aggregate. These positions are worked intensively. The Consultants have been shown side-scan and echo-sounder records which demonstrate that this can produce a pock-marked location (anchor pits) or possibly a large scale shallow trench in the seabed as a result of trailer dredging, (the width of which may be about 100m when mainchain Decca is used for position fixing - see Section 10.2.3).

As the workable deposit thins, high-spots in the underlying deposit (eg sand, chalk, clay) are exposed, thus posing the danger of cargo contamination. When this contamination exceeds a certain critical level the location or course is considered 'worked out', and a new one sought. To some extent such bad spots along trailer courses can be avoided, thus accommodating the geometry of the workable deposit; courses are however normally orientated along the prevailing axes of tidal flow.

It must be stressed that no gravel dredging company dares to allow its cargoes to be more than fractionally contaminated with sub-strata materials, otherwise the landed material could be financially worthless. Additional costs could also be involved in removing contaminated stockpiles from the wharf if the material cannot be processed by on-shore plant to meet the requirements imposed in BS 882:1983 for concreting aggregates.

These observations have suggested a number of questions which the Consultants believe should be considered in developing future strategies for the management of the working of marine aggregate resources.

i) Could trailer dredging courses be better related to isopachytes (lines of equal deposit thickness) and zones of equal deposit quality? Areas could then be defined where dredging could take place to a predetermined depth without fear of contamination. 'Awkward' zones in the vicinity of outcrops of the underlying deposit could be avoided, and areas of localised thickening of deposits could be designated as anchor dredging zones. One of the main problems faced in implementing such ideas is the degree to which trailer dredging courses are controlled at the present time by the direction of the tide (DoE, 1986). Aggregate dredging across the tide is not normal practice, due to the dangers during rough weather of damaging the dredge-pipe in circumstances when it tends to move underneath the ship. However, dredging across the tide is used in civil engineering projects such as trench digging, and has on occasions been employed by aggregate dredgers. Should the benefits of relating dredging courses to geology rather than tidal flow become widely recognised, it is likely that state-of-the-art developments in dredge-pipe articulation hardware (Section 5.2.3), and pipe and ship computer-controlled positioning systems (Section 5.3.1) will significantly reduce the level of risk to dredging gear.

Trailer dredge pipes and draghead

ii) Does trailer dredging as practised produce wide trenches on the seabed? If so, does this leave intervening ridges of good quality material that are difficult to dredge because of the high risk of contamination by underlying material when the pipe moves into the adjacent 'dredged out' trench? The Consultants believe that this may be the case and suggest that the physical changes on the seabed resulting from present day trailer dredging practices should be the subject of study (a project which could usefully be funded by Central Government). Further, the Consultants suggest that any tendency towards trenching rather than planing can be alleviated by the use of high accuracy position-fixing systems (Section 6.2.4), as courses navigated can be varied by a matter of a few tens of metres on a regular basis so ensuring an even removal of the seabed; in some instances courses intersecting at shallow angles can be superimposed over a period of time, further reducing the chances of missing areas of deposit (Figure 6.4). Tracks can still be broadly aligned to the direction of strongest tidal flow.

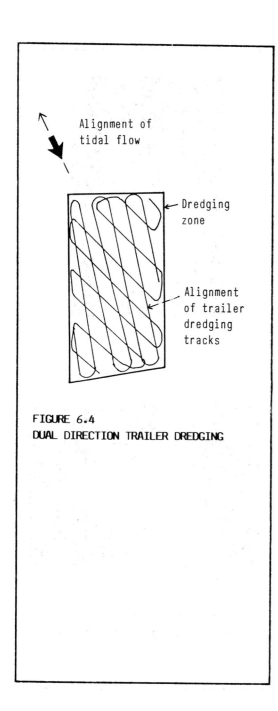

FIGURE 6.4
DUAL DIRECTION TRAILER DREDGING

The Consultants believe that many benefits should accrue from the adoption of this type of more controlled seabed mining.

Of immediate relevance to the operators are:

i) Contamination as a result of exposure of underlying layers may be reduced.

ii) A higher percentage of recoverable reserves from the ground could result.

Wider benefits include:

iii) The ability to predict sea-floor level changes more accurately, thus enabling more realistic modelling of coastal stability implications (Section 9.4).

iv) The return of a relatively smooth sea-floor of uniform substrate type to the fishing industry, with localised anchor dredging areas charted for fishery avoidance (Section 10.4.3).

v) Improved positioning accuracy combined with better knowledge of the geology and geometry of deposits could allow dredging vessels to respond more accurately to the day-to-day grade requirements of the shore plant.

It should be pointed out that dredging accuracy could never be constantly maintained, as would be attempted for example in civil engineering projects; the need to dredge in all weathers and within a turnround cycle could mean that accuracy would often be relaxed. On a statistical basis however the end result would be a far more controlled excavation of the seabed than presently occurs.

The prerequisites for the development of such methods of mining aggregates are accurate position fixing (Section 6.2.4), extensive and accurate prospecting data (as described in Section 6.2) and agreement between the Crown Estate Commissioners and Industry to implement the methods whenever practical.

6.3.2   Prevention of sterilisation of reserves by screened out material ("over sanding").

The screening-out at sea of a percentage of the sand content of aggregate 'as raised' is now common practice in the southern North Sea (DoE, 1986). This is necessary to provide a readily saleable cargo with the optimum ratio of sand to gravel (about 40% sand and 60% gravel, by weight).

The screened-out sand is returned to the seabed within the dredging licence area. This can lead to a gradual increase in the percentage of sand within the upper layers of the seabed, although in some instances this may be temporary due to the reworking capabilities of tides and waves. This is particularly a problem for trailer dredging, because the dredge head travels over the seabed surface, removing the surface layers. As the percentage of sand increases, more dredging effort is needed, with a higher percentage being screened out, to achieve a saleable cargo. Eventually, the area may become uneconomic to work, perhaps with further valuable reserves of saleable quality sterilised below a cap of unsaleable sand.

The solution to the problem of excess sand would ideally come from the marketing side, but there appears to be only limited ways forward here. The Industry is aware of possible solutions, such as the use of surplus sand for beach replenishment and fill, which is one answer that is becoming more common, and will clearly help to relieve the situation. Also there could be an increasing reliance on additional marine sand to replace exhausted land reserves, particularly if the alternative sources involve crushed rock as the coarse aggregate component in concrete. In the remaining instances a solution must be sought at the dredging end of the operation to a problem which, in some grounds, may be progressively sterilising valuable deposits.

A similar problem is faced in a planned metaliferous placer mining project in the UK where 50% rejection is proposed in a multi-pass trailer dredging operation. To enable working of the lower ore-bearing layers a novel tidal-mining system has been suggested, which should be put to the test in coming years. In the Consultants' opinion the system is equally applicable to aggregate dredging in those instances where it is necessary to reject large quantities of sand.

Screening and rejection of sand on an East Coast ground; night-time dredging

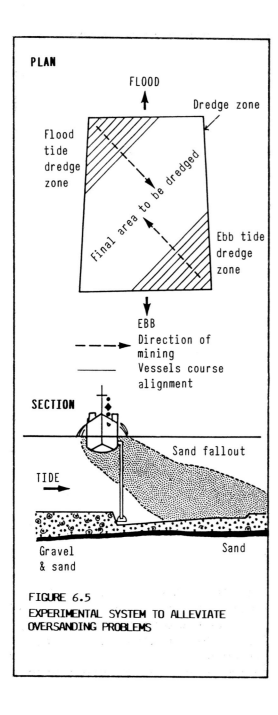

FIGURE 6.5
EXPERIMENTAL SYSTEM TO ALLEVIATE OVERSANDING PROBLEMS

The details of the proposals are outlined in Figure 6.5. The scheme utilises the ability of the tide to transport rejected sand, and will therefore be most efficient in areas where there are not extended periods of slack water and where tidal flow patterns are approximately rectilinear. A flood tide and an ebb tide dredging area are designated, and each is utilised according to the time relative to high water at which dredging commences. Dredging takes place at a slight angle to the tide. Using an accurate position fixing system a 'furrow', extending the full depth of the deposit, is progressively worked out towards the central area of the dredging ground. Rejected sand is carried by the tide to one side of the ship and effectively 'back-fills' the worked area. Sand settles at rates of between 3mm/s (medium sand) to 20mm/s (2mm granules); taking a value of 5mm/s (for coarse sand of which the bulk of the reject is thought to be composed), in 25m of water the back-fill will be carried 40m from the ship per knot of tide.

A complication arises where tidal streams are strong enough to cause a dispersion of rejected sand once it reaccumulates on the seabed. In areas of strong residual sand transport an 'over-sanding' problem may not exist, as sand is carried from the dredging area. Where waves, currents and tidal streams are only strong enough to cause a slow dispersion of sand by 'bed-load' transport, dredging would have to be restricted to the downstream end of the bed-load pathway, and could only take place over one half of the tide.

Applicability of the system is clearly not universal and will depend upon local conditions (sea-state, tides, geology) and vessel capabilities. (the problems associated with dredging across the tide have been discussed in Section 6.3.1). It is however an idea worthy of consideration within future mining strategies.

6.3.3   Progress monitoring

The mining management system proposed in the previous section would require a periodic assessment of the development of a dredging ground and of the depletion of its reserves.

In the Consultants' view, with accurate base-line prospecting data on the geometry of the deposit, and detailed records from within each dredging area of extraction positions, tonnages and aggregate quality, annual or biennial bathymetric (echo-sounding and side-scan) surveys are all that would be necessary to complete the development picture. Such data would provide:

i) A check on dredging progress and identification of areas 'missed'.

ii) Assessment of the importance of any natural replenishment mechanisms, and also

iii) provide information which could be used in relation to any new problems which have arisen concerning coastal stability or fisheries since the issue of the licence.

The Consultants understand that some companies are starting to implement baseline surveys for future progress monitoring programmes.

6.3.4   Dredging control

In the Consultants' view the restriction of dredging activity to licensed areas relies largely on trust between the Crown Estate Commissioners and the dredging companies; and between the dredging companies and individual masters of vessels.

There have been a few cases where penalties have been imposed when dredging off station has been observed and conclusively proved (DoE, 1986). Although illegal dredging is generally thought to be essentially a problem of the past, the Consultants strongly recommend the introduction of electronic activity recording systems. As allegations of dredging off-station still (usually mistakenly) occur, this step would short-circuit the present time-consuming and expensive procedure which has to follow every complaint, and safeguard the dredging industry against false accusations. It is also hoped that this action would, for the future, form the basis of increased trust and better relationships between the dredging industry and other groups with marine environmental interests.

The sealed recording systems located on each ship would constantly monitor the ships's position, and operation parameters diagnostic of whether dredging is taking place. In the Consultants' view it would be a great pity if the installation of position-tracking equipment* was enforced in isolation of the possible future developments of mining management systems described in the preceding sections.

If such a system were to be introduced the activity of all aggregate dredgers in UK waters should be monitored. This would of course require the same monitoring system to be installed on any dredger registered abroad and licensed to dredge in UK waters. However, monitoring the movements of unlicensed foreign dredgers and of UK registered navigation dredgers would continue to be noted by MAFF fisheries spotter planes and UK fisheries patrol vessels.

---

*It is unlikely that mainchain Decca, only accurate to the nearest 0.5 nautical mile under certain conditions, will be suitable for dredging monitoring.

It must be recorded however that the aggregate dredging industry, as represented by SAGA and BACMI, strongly opposes the introduction of position monitoring systems, as this would amount to dredging vessels being treated differently from merchant and fishing vessels generally. The Industry further points out that no sealed recording systems have yet been developed, and that development costs (not yet fully assessed) would need to be borne by Central Government.

mv El Flamingo (see Section 7.1.2)

Photo courtesy of A Duncan, Ship Photographer, Gravesend, Kent

THE DEVELOPMENT OF PROCESSING AND QUALITY CONTROL

7

CONTENTS :

Aggregate Processing

　Shore processing
　Shipboard processing

Quality Control

　Standards development
　Combatting user prejudice

7.1 AGGREGATE PROCESSING

7.1.1 Shore processing

The processing of marine sand and gravel will follow closely any technological developments in the processing of land-won sand and gravel. During the post-war years, with an increasing demand for aggregates, plant manufacturers have improved the design and operating efficiency of their machinery (Section 11.3.2).

In establishing new wharves mineral operators will attempt to maximise the value of their products by seeking permissions for concrete batching plants and/or asphalt plants (Section 11.3.3).

7.1.2 Shipboard processing

A dredger was designed for onboard processing in 1967 but as a processing unit it was uneconomic because of existing investment in onshore plant coupled with the fact that customers were unwilling to pay the premium for ready processed aggregate. Dredged material was initially screened @ 140mm down to 10mm at sea and then pumped ashore over 20mm screens.

Around 1970 the tanker mv British Defender was converted into a floating suction hopper dredger/processing plant and re-named mv El Flamingo (5845 GRT, - photograph in Chapter 5). After a short period of service in the Thames Estuary she was found to be uneconomic. This was because of a large maintenance crew (36 men), and the constant clogging of screens by sand which resulted in very high running costs. Had the Maplin Air/Seaport project gone ahead the El Flamingo might have been used successfully. The vessel, which was not fully utilised, was sold in 1977 and laid up before being scrapped in 1984.

Because of the cost and operational difficulties experienced with shipboard processing, the marine aggregates industry now limits its processing to single deck screens

Shells and mud balls screened from sand at sea

(DoE,1986) to approximate shoreside requirements (either for direct sales or further processing).

7.2 QUALITY CONTROL

7.2.1 Standards development

British Standards affecting sand and gravel are being revised, updated or in some cases introduced for the first time. They may affect limits for particular values or describe new methods for carrying out laboratory analyses. A number of standards relate to aggregate testing. Probably the most important standard in terms of overall limits for aggregates used in concrete is BS 882: 1983 which sets out, among other matters, specifications for grading requirements, shell and chloride content.

Grading requirements: Reference is made in Section 3.1.1 to the use of different scales for measuring particle sizes. The aggregates industry use the specifications given in BS 882: 1983 (Aggregates from natural sources for concrete) which are not followed by soil mechanics or academic geologists who use other British Standards (see below).

Care must therefore be taken to quote the appropriate British Standard when applying descriptions to aggregates. The most important difference is the classification of coarse and fine aggregate. BS 882 sets this boundary at 5mm whereas other Standards (eg BS 1377: 1975 - Methods of test for Soils for civil engineering purposes, and BS 5930: 1981 - Code of Practice for Site investigations) set this at 2mm.

Shells in dredged aggregates are not seen as a problem by the industry: limits are given in BS 882: 1983 and screening/processing removes most shells with only fragments remaining. A fuller description of shell related problems is given in DoE, 1986.

Chlorides: The fears concerning salt contamination arise largely from the possible corrosion of embedded metal in concrete. This risk particularly is important to any reinforced concrete component and especially to pre- and post-tensioned concrete. The sulphate resistance of SRPC (Sulphate Resistant Portland Cement) is also thought to be reduced. There is also some concern that there is link between salt (as a de-icer on roads) and possible affects from alkali silica reactions (see below). The source of these chlorides was thought to have been solely from the sea and therefore dredged aggregates have been looked upon with suspicion by some specifiers and users (Section 7.2.2). However, chlorides are also present in beach materials and some inland aggregates.

Another source of chloride ion was in $CaCl_2$ where it was used as an admixture in concrete to accelerate both the setting time and the rate of gain of strength. The use of $CaCl_2$ has been banned since 1985 with the publication of BS 8110: Part 1 (Code of Practice for design and construction).

It would appear therefore that the probability of excess chlorides contained in concrete is now greatly reduced following the recent banning of $CaCl_2$. This should assist in combatting any user-prejudice (Section 7.2.2).

BS 882: 1983, (Table 8) sets the limits to the maximum total chloride content of **combined aggregate** and is the measure normally followed by the aggregate producers. This is expressed as a percentage of chloride ion by mass:

> For pre-stressed concrete and steam-cured concrete this is set at 0.02%; for concrete made with cement complying with BS 4027 (Specification for sulphate-resisting Portland cement) or BS 4248 (Supersulphated cement) this is set at 0.04%; and for concrete containing embedded metal and made with cement complying with BS 12 (Specification for ordinary and rapid hardening Portland cement) this is set at 0.06% for 95% of test results, with no result greater than 0.08%.

BS 882: 1983 (Table 8) also points out that the limits stated are derived from those given in CP 110 : Part 1, now replaced by BS 8110: Part 1: 1985, which refers to the total chloride content of **concrete.**

BS 8110 is followed by concrete manufacturers because the non-aggregate constituents of concrete may also contain chlorides, eg in any admixture, in the mixing water, and in the cement itself (Martin, 1984).

BS 882: 1983 draws attention to the fact that "marine aggregates and some inland aggregates contain chlorides. Both should be selected carefully and may need efficient washing to achieve the limit required for use in pre-stressed concrete". No limit of chloride is set for coarse and fine aggregate used in concrete containing no embedded metal, ie plain (mass) concrete.

The conclusion therefore is that **compliance of the aggregate with BS 882 does not guarantee compliance with BS 8110.** However, **this does not imply that marine aggregates need be of an inferior quality to land-won sand and gravel.** Provided the chloride levels in the hardened concrete are below the stipulated maxima and provided normal criteria for concrete quality and depth of cover are applied, no significant **additional** corrosion risk should be encountered by the use of marine dredged aggregate (W H Gutt, BRE, pers. comm.).

Settlement of fines

Carbonation of chloride-bearing concrete substantially increases corrosion risk over that prevailing in carbonated but chloride-free concrete (W H Gutt, BRE, pers. comm.).

A more detailed discussion on chloride specifications (the chloride controversy) is contained in DoE, 1986.

Alkali - aggregate reactions in concrete: In recent years there has been increasing recognition of the fact that alkalies can react with certain types of aggregate in concrete structures - a phenomenon known as the alkali-aggregate reaction (AAR). Although this problem is believed to be of limited occurrence, when reactions do appear they can be severe and expensive to remedy. There are various types of reactions but only that known as the alkali-silica reaction (ASR) has been identified in the UK. The chemistry of ASRs is not yet fully understood.

There is no British Standard on alkali-aggregate reaction and the reader's attention is directed to BRE Digest 258 'Alkali aggregate reactions in concrete' (1982) and the guidance notes given in the Report of a Working Party on 'Minimising the Risk of Alkali-Silica Reaction (the 'Hawkins Report'), September 1983. A draft technical updating report "Minimising Risk of alkali-silica reaction - Guidance Notes and Model Specification Clauses" was published by the Concrete Society in Autumn 1985.

A draft of BS 812: 123 Testing Aggregates : Method for determination of alkali-silica reaction is to be published in Spring 1986. Troy (1985) has suggested that at present **the cement industry should take the initiative in cutting down alkali where the main problems arise.**

ASR occurs when alkalies, normally present in Portland cement, react with some forms of silica in aggregate to form a gel. This gel absorbs water and hydroxide ions (originally released when water was added to the cement), swells and exerts a pressure which can crack concrete: known as an expansive effect.

The important point with ASR is that damage only occurs if **all** the following factors are present:

i) moisture (mix water used to make fresh concrete workable may be sufficient),

ii) sufficient alkali (usually from cement), and

iii) a critical amount of reactive silica in the aggregate (nearly all sands and most gravels contain reactive silica).

There has also been recent concern expressed among engineers that road salts may induce AAR on the surface of concrete which then cracks, allowing deeper penetration of salts, further cracking and renewed action (Byrd, 1985). Byrd commented that the problem of aggregates contaminated by chlorides which add to the reactive potential of high alkali cements is now seen as more serious than first thought. However, Jardine (1985) confirmed that lower chloride levels are now being demanded both in the concrete mix and in the aggregates, whether they are sea-dredged or land-won. Jardine stressed that all producers must know the chloride contents of their materials and should consider the need to reduce them.

Jardine (op.cit) pointed out that there are no examples of ASR where marine aggregates have been used in the east and south of England. Many cases have been reported in the South Devon area where high-alkali cement was probably used and a number of cases have been reported in the rest of Devon, parts of Cornwall, Somerset, Avon, Gloucestershire, Gwent and in the West Midlands, Staffordshire, Leicestershire, Derbyshire and Nottinghamshire. Many of these involve the use of Trent Valley gravels. Isolated cases have been reported in the London area, Suffolk and Strathclyde. In recent years the cement industry in Devon has reduced the alkali content in its products by blending with bastfurnace slag (A T Corrish, BCI, pers. comm.).

There is increasing concern about AARs especially when buildings and motorways are suspected of being affected. Both the popular and scientific Press have given coverage to this problem. However, as Jardine (op. cit) implies, the significance of ASR to the marine sand and gravel industry is negligible, particularly when the producers satisfy the chloride limits now demanded and know what potentially reactive combinations of different aggregates to avoid or can prescribe a safer mix.

Sims (1983) reported on a series of tests carried out on concrete mixes prepared with sand and gravel dredged from the North Sea (Cross Sands Bank) in combination with crushed Carboniferous Limestone from the Mendips and 'low-alkali' Portland cement from Northfleet, North Kent, and 'high-alkali' cement from Plymstock, South Devon. Two types of granulated blastfurnace slag were also used, one freshwater quenched and the other seawater quenched. The result showed that the dredged flint aggregate has relatively low alkali-reactivity potential, even when blended with limestone aggregate and/or high-alkali cement. However, Nixon and Bollinghaus (1983) reported that there was greater reactivity of aggregates dredged from the Thames Estuary, South Coast and Bristol Channel areas together with sand and gravels from inland sources (Thames and Trent Valleys) which all contain potentially reactive flints or cherts.

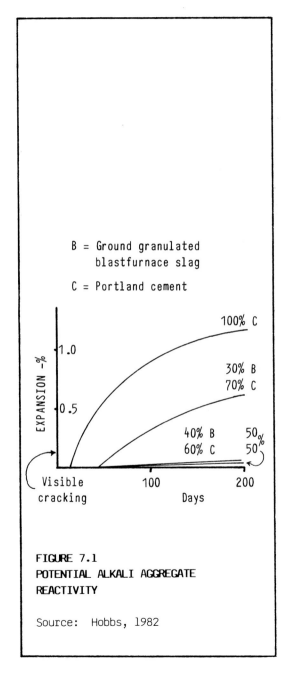

FIGURE 7.1
POTENTIAL ALKALI AGGREGATE REACTIVITY

Source: Hobbs, 1982

Where ASR may occur measures can be taken to minimise the risk of damage which include the avoidance of high alkali cement or the addition of pulverised fuel ash or ground granulated blastfurnace slag (Figure 7.1).

While the sand and gravel industry must not be complacent about ASR **continued testing would appear expedient for all aggregate producers.** Jardine (1985) felt it necessary for operators to determine the type and proportion of reactive silica and the mix proportions that give the greatest assurance of safety.

Shrinkage in concrete: Certain aggregates have been found to expand when wet and shrink on drying which can cause serious defects in concrete structures. This phenomenon is particularly prevalent in West-Central Scotland. British Standards are to be published during 1986. The areas where marine sand and gravel are currently dredged do not appear to have given rise to any drying shrinkage problems. Further details are contained in DoE, 1986.

7.2.2   Combatting user prejudice

This section is an account of overall impressions gained by the Consultants after talking to a number of people who represent the suppliers of marine aggregates and the specifiers of what aggregate (not) to use. Particular views are credited in appropriate places.

Unfortunately for the marine aggregates industry it would appear that there are many architects and surveyors, in both the private and public sectors, who are ignorant of the facts relating to chloride ion concentrations in concrete (DoE, 1986). Some have considered the evidence but are still not prepared to recommend the use of marine aggregates, deciding to continue specifying the use of land-won sand and gravel. Similar arguments are often advanced for shells and ASR.

The effects of this are often beyond the control of the Industry and cases have been reported where established concrete batching plants have had to switch supplies of aggregate from marine to land sources in order to maintain an operating presence in an area (J S Ornsby, RMC, pers. comm.).

The sand and gravel industry indicate that user prejudice appears to have been more widespread ten to fifteen years ago. Today this seems to be patchy across the country, even within a large area such as Greater London. Some wharves experience no apparent problems with sales whereas others find difficulty with certain organisations. In 1984 the GLC carried out a survey into marine dredged aggregates in public works contracts within Greater London. Of the 27 authorities who responded one third of the respondents discriminate against the use of marine aggregates in some way or another.

The GLC survey revealed that discrimination can take several forms :

blanket prohibition [notwithstanding the fact that BS 882: 1983 sets limits on both shell and chloride content for aggregate generally and **not** type (source) of aggregate] - some **inland** sources and estuarine deposits display high shell and chloride contents;

reluctance to use marine aggregates in 'waterproof construction' because of alleged risks of permeability caused by shell content [many other aggregates contain substantial amounts of flaky materials];

alleged problems over calcium chloride [misplaced judgement since $CaCl_2$ is used as an admixture to accelerate the setting of concrete];

The following comments typify the attitude of some Authorities on the acceptance of marine aggregates:

'in certain circumstances, with the express approval of the Architect'; or

'only "natural" aggregate in site mixed and ready mixed concrete supplies'; or

'do not use for site mixed concrete'; or

'reluctant to use unless completely satisfied that the requirements of the British Standard could be met'; or

'reluctant to introduce a further hazard into the durability of concrete due to salts and perhaps marine life'.

It would appear that the GLC survey indicates a widespread level of ignorance on the properties of marine dredged aggregate compared with other sources of aggregate and the levels set down by the BS 882: 1983. It is perhaps instructive here to draw to the attention of those who have 'reservations' on the use of marine aggregates the definition of what a Standard should be, which is given in BS 0: Part 1: 1981 -

"A technical specification or other document available to the public drawn up with the co-operation and consensus of all interests affected by it, based on the consolidated results of science, technology and experience, aimed at the promotion of optimum community benefits and approved by a body recognised on the national, regional or international level".

If the users of British Standards are therefore dissatisfied they should direct their criticism to the respective BSI Committee Secretaries.

The preparation of a Standard will often take a considerable period of time to finalise in order to evaluate fully the technical and scientific arguments and to arrive at procedures and limits/levels which are acceptable to both supplier and user.

From the survey results collated by the GLC it is clear that it is **not** BS 882: 1983 which is at fault but those who misplace or misinterprete the Standard and the reference to marine aggregates contained therein.

The use of a common standard for chlorides in an area like London is obvious. Through the Seminar of London Borough Structural Engineers (SEMSEL) it is proposed that all boroughs adopt the chloride levels in concrete which are specified in CP 110 for civil engineering and highway works (D Williams, GLC, pers. comm.). While this is a step towards adoption of a unified standard it still does not overcome the different standards at present operated by the GLC's Transportation and Development Department with those of the Architect's Department.

Some local authority architects do not have a standard specification for concrete and tend to use the NBS system (Martin, 1984) which allows the specifier to select clauses appropriate to the works. One of the optional clauses is "do not use marine (sea dredged) aggregates". Whilst that clause is normally deleted external consulting engineers sometimes prefer to see it retained (V A Veck, Hants C C, pers. comm.). Other clauses allow the use of sea dredged aggregates with limitations.

Although marine sand and gravel per se is not injurious in concrete the fact that it is dredged from the sea which contains NaCl creates unjustified fears. Marine aggregate is geologically similar to sand and gravel excavated from land-won sources (R A Fox, RMC, pers. comm.) and is usually washed before sale to remove surface films of sea water : these two facts appear to be overlooked by some customers.

When compared with marine aggregates, there is little requirement by customers for laboratory testing of aggregates from land-won sources even though it may have chloride concentrations and may also be receptive to ASR. **More testing of land-won aggregates seems likely in the future** (I Sims, Sandberg, pers. comm.).

Given the limits set out in Table 8 of BS 882: 1983 for total chloride ion concentrations, it is therefore difficult to see how the marine industry can overcome any prejudices. Some operators with wharves drawing water from tidal rivers have changed or are changing to a mains supply. This is not because of any chloride problems in the final product. It is solely to ensure a continual supply of fresh make-up washwater for processing in times of extreme drought (Section 11.3.2).

The Consultants understand that in some areas of the UK, for example North East England, marine aggregates are regarded as superior concreting products to local land-won sand and gravel which suffers to varying degrees with contamination by coal fragments and other deleterious rocks incorporated into the deposits form the Ice Age. Thus in some locations there may actually be a preference to use marine aggregates.

INTERACTION WITHIN THE COASTAL ENVIRONMENT

PROBLEMS OF PHYSICAL AVOIDANCE

8

CONTENTS:

Interference with navigation
- Statutory responsibilities
- Consultation procedure
- Contraints on dredging operations.

Zones designated for alternative uses
- Waste disposal areas
- Marine nature reserves
- Pipelines and cables

8.1 INTERFERENCE WITH NAVIGATION

8.1.1 Statutory responsibilities

The Department of Transport (DTp) is the custodian of the public right of navigation and is reponsible for safety at sea. Any proposed operation, including aggregate dredging, which is likely to obstruct or endanger navigation must be scrutinised in this respect. The statutory control exercised by the Department is described in Section 1.3.3.

All dredging vessels are also required to comply with the Collision Regulations (Ship and Seaplanes on the Water) and Signals of Distress (Ships) Order 1965 (SI 1525). Whereas normal rules of the road apply to dredgers steaming or loading at anchor, a trailer dredger is considered unable to avoid approaching vessels whilst loading and must carry marks and lights accordingly (Figure 8.1).

8.1.2 Consultative procedure

In the first instance DTp is approached directly by the dredging company for a consent to prospect. Details of the CEC licensed prospecting area and survey methods are supplied. DTp consults the navigation bodies likely to be affected by the proposals (see below) and depending on advice received issues a consent with any conditions imposed. The consent normally expires after two years.

Once a full licence application is submitted by a company, DTp is consulted as part of the **Government View** procedure (Section 1.3). Details of the area to be worked, estimated annual quantity of material to be dredged and extraction methods are forwarded to navigation bodies which may be affected, including:

The General Council of British Shipping, for consideration of interaction with commercial shipping interests.

The Hydrographer of the Navy, relative to potential effects on the regime of the seabed (shipping channels).

General Lighthouse Authority, with regard to the protection of sea marks (buoys, fixed structures, lightships).

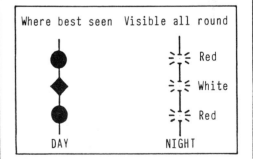

FIGURE 8.1
LIGHTS AND SHAPES IDENTIFYING DREDGER OPERATION

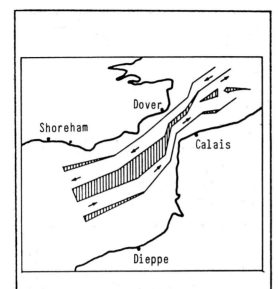

FIGURE 8.2
STRAITS OF DOVER
IMO Traffic separation scheme.

*Criteria used in the definition of a busy waterway are the density of shipping and the degree to which this shipping is obliged to follow particular routes by reason of the draught.

Harbour Authority, concerning interaction with shipping where a port area of jurisdiction extends into proposed dredging area.
Local Officer, (ie coastguard or nautical surveyor), for general safety considerations if close inshore and there is no local port.
Royal Yachting Association, for consideration of interference with leisure sailing activities.
Ministry of Defence (Navy), concerning RN and NATO exercise areas.
Pilotage Authorities (normally Trinity House Pilots) for pilotage considerations.

Responses are collated by DTp, who intimate to DoE whether consent will be forthcoming and what restrictive conditions would need to be imposed. On attainment of a CEC Production Licence, the company involved formally applies to DTp for consent, which is issued on a renewable three yearly basis. Consent is subject to there being no variation from the proposed dredging programme without prior consultation, and the proviso that the consent can be withdrawn at any time by the Secretary of State.

8.1.3   Constraints on dredging operations

In past years four licence applications have been refused on navigation grounds, and another subject to reduction in working area. Three of the refusals were in the Bristol Channel. Generally the attitude of navigation authorities towards aggregate dredging appears to have relaxed. In early days it was often stipulated that vessels must anchor to dredge in busy waterways*, whereas nowadays conditions usually relate to the number of vessels working at any one time and the direction of trailer dredging. The only major issue which has arisen since 1973 relates to the Port of London approaches, where dredging in the Sunk/Black Deep area was opposed. The DTp have recently reconsidered this case and withdrawn their objections.

Only two types of area now persist where DTp would be unlikely to accept dredging for aggregates under any circumstances:

(i) **Traffic lanes** of internationally (IMO) adopted traffic separation schemes, such as in the Dover Straits (Figure 8.2).

(ii) **Protection areas around navigation marks.** Typically dredging is not permitted within 0.5n miles radius of a floating aid to navigation but the distance may be increased around major aids (eg lightships).

The interrelationship of navigation and dredging interests is maintained on a consultative basis, and fortunately the exchange is such that both sides are well aware of each others problems. The operation of aggregate dredgers close to major shipping routes was an intial cause for concern and it is presumably due to the excellent safety record of the past 10-20 years, combined with improvements in position fixing and radar techniques, that relaxation of opposition has taken place. **In the Consultants' view it is probably true to say therefore that the future maintenance of minimal DTp constraints on aggregate dredging depends largely upon a continued record of operational safety.**

8.2 ZONES DESIGNATED FOR ALTERNATIVE USES

8.2.1 Waste disposal areas

Industrial wastes and sewage sludge. At least five of the thirteen offshore dumping grounds currently receiving this type of waste are in seabed areas of sands and gravels which might at some stage be considered as aggregate reserves. Futhermore, in many instances the zone affected by solid waste fallout far exceeds the designated dumping area (Norton et al,1984), greatly increasing the area of potential conflicting use. At the present there is no objection to gravel extraction within certain dumping grounds, as exemplified by the area off the River Humber (Section 10.2.2); this situation might conceivably change in the future.

Harbour-dredgings dumpsites. Some seventy harbour spoil disposal sites are used in UK waters, usually lying in 20-30m water depth (Murray and Norton,1979). Often established during the last century, and subject to many years of continual use, the despoliation by mud, dock debris and capital dredgings is commonly extensive, effectively burying any underlying aggregate reserves.

Unfortunately, no data are published quantifying the total area of waste disposal grounds (see Figure 8.6).

8.2.2 Marine nature reserves

A distinct possibility for the future is the setting up of offshore nature reserves under the auspices of the Nature Conservancy Council. These areas, of many square nautical miles extent, will be effectively closed to dredging (this subject is discussed further in Section 10.1.2).

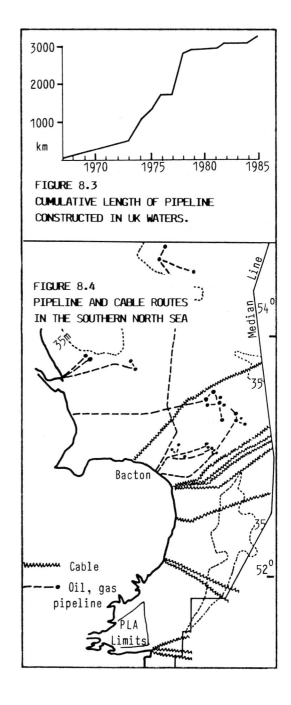

FIGURE 8.3
CUMULATIVE LENGTH OF PIPELINE CONSTRUCTED IN UK WATERS.

FIGURE 8.4
PIPELINE AND CABLE ROUTES IN THE SOUTHERN NORTH SEA

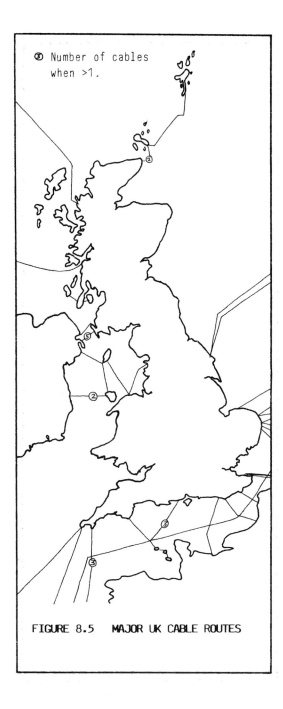

FIGURE 8.5   MAJOR UK CABLE ROUTES

8.2.3   Pipelines and cables

Because of the high cost of emplacing conduit (necessitating shortest most direct routeing) and the need for an indisturbed service life (resulting in the careful selection of seabed routes) there is only limited scope for deflecting new pipelines and cables around existing or potential aggregate dredging areas. As dredging is not allowed close to pipes or cables, this precedent effectively excludes large areas of the continental shelf from aggregate mining.

Oil and gas pipelines, well-heads and platforms. Since the mid-sixties many oil and gas production structures have been sited on the UK continental shelf, including hundreds of kilometres of pipeline (Figure 8.3). The southern North Sea is of particular import to the dredging industry, where there are seven major gas fields and linking pipelines (Figure 8.4). These contain approximately 850km of pipeline and 36 well-heads and platforms.

By international law there is a safety of 500m radius surrounding oil and gas structures, in which navigation is prohibited without the permission of the operator. No similar legal enforcement applies to pipelines, which are normally buried and hence constitute no danger to navigation. It is however an offence to wilfully damge a pipeline, and the Department of Energy recommends a 'safety corridor' of 500m on either side of the pipe within which dredging should be avoided. Closer limits are not considered practical due to the position fixing systems relied on by aggregate dredgers (mainchain Decca). In some instances CEC have entered into agreements to define pipeline and cable (see below) safety corridors.

Safety zones and corridors presently cover some 900km² of the seabed in the southern North Sea (Figure 8.4), and new pipelines are planned between Bacton and the Indefatigable Field for the next few years.

Cables. Some 30 cables cross the continental shelf of the English Channel and southern North Sea, there being more than twice as many routes within all UK waters. Most are telephone cables, the notable exception being the Dungeness-Boulogne and Folkestone-Sangatte electricity lines, the latter in the process of being laid. Major cable routes are shown in Figures 8.4 and 8.5. In the southern North Sea alone (UK sector) there is approximately 1000km of active cable.

As with pipelines, there is no legal enforcement of protection zones along cables, although again it is an offence to wilfully cause damage. The Marine Divison of British Telecom recommends an exclusion zone relating to seabed activities for one nautical mile (1.8km) on either side of the cable. This corridor has to be wider than for pipelines for two reasons:

(i) Prior to 1980 cables were laid using mainchain Decca position fixing, and consequently their precise location is unknown.
(ii) The cables are not buried, and may move slightly under tidal action or be snagged and towed out of position by fishing vessels.

In consequence of the 3.6km wide protection zone, some 3,600km² of seabed in the UK sector of the southern North Sea are barred to aggregate dredging.

In order of magnitude figures, one tenth of the seabed of the southern North Sea that is within the reach of modern aggregate dredgers (~15-35m) is not accessible due to the presence of pipelines or cables. The situation is somewhat better in the English Channel and off other coasts. There is little possibility of a reduction in areas sterilised in this way, whereas the situation could worsen if aggregate dredging were ever held responsible for damage to pipes or cables (possibly another argument for use of more accurate position fixing systems on dredgers). The density of pipes and cables on the seabed could also increase. The point should be made that bottom fishing is also excluded from pipe and cable protection zones, thus increasing pressures of 'inter-user' conflict in the remaining areas of the seabed (Section 10.4).

FIGURE 8.6
MAJOR UK WASTE DISPOSAL GROUNDS

Source: MAFF

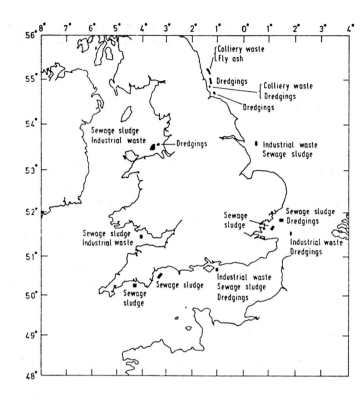

SHORELINE STABILITY

9

CONTENTS:

Responsibility of shoreline protection
  Organisation
  The extent of coastal erosion problems in the UK.

The dynamics of beach and nearshore environments
  Waves at the shoreline
  Beach characteristics
  Beach and nearshore sediment exchange.
  The supply of beach sediment from offshore.
  Man-induced coastal change

Dredging related research
  The history of research
  Research results
  Research and licensing

Shoreline considerations in future perspective

9.1 RESPONSIBILITY FOR SHORELINE PROTECTION

9.1.1 Organisation

For historic reasons local responsibility for the stability of UK coasts is divided in two. **Coast protection,** a term which has come to mean work carried out to protect land against erosion, is primarily the responsibility of maritime local authorities, with some involvement of other bodies such as port authorities, county councils (as highway authorities), or nationalised industry. These authorities have statutory powers under the Coast Protection Act 1949 to carry out works, but these are permissive. There is no obligation for a coast protection authority to carry out such work although the Minister for Agriculture, Fisheries & Food retains default powers which can be invoked if he considers the authority to be failing in its duty. Joint Coastal Protection Committees are presently being considered (Boxer,1985) to provide continuity of effort along coasts under the jurisdiction of several authorities.

Sea defences are works for the protection of low-lying land against flooding. Originally arising from land reclamation programmes, defences are maintained under the Land Drainage Acts. Sea defences are the responsibility of the Regional Water Authorities.

Ultimate responsibility for protection lies with MAFF, DAFS and the Welsh Office, and these authorities administer grant aid. As financial benefits of coastal proctection schemes are required to outweigh construction costs, grant-aided works are effectively limited to coastlines with substantial development.

A Green Paper is currently before Parliament to consider the transfer of all shoreline protection responsibilities under the single umbrella of the Regional Water Authorities.

Protection and defence works are undertaken by engineers, both consulting and in the employ of the statutory authorities, and have traditionally involved constructional projects such as sea walls and groynes. In the last 25 years more consideration has been given to the understanding and management of nearshore and intertidal sediment regimes and the use of beach replenishment.

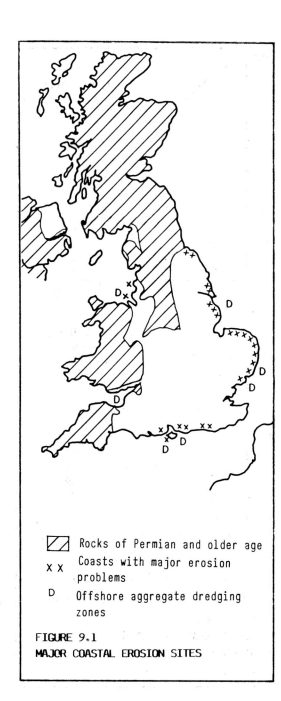

FIGURE 9.1
MAJOR COASTAL EROSION SITES

9.1.2 The extent of coastal erosion problems in the United Kingdom

Coastal erosion may be found along sections of almost any coast, resulting from a variety of causes including natural local cyclic phenomena, worldwide slight climatic and sea-level changes, or the interference of man. Beyond these local effects it is possible to define, in terms of exposure to wave attack and the properties of the coast-forming rocks, areas of persistent long-term erosion (Figure 9.1). Most of the rocks prior to the Permian era are of a hard resistant nature, and form coast-stabilising strongpoints which minimise erosion problems in South West England, Wales, Northern England, Scotland and Northern Ireland. The remaining areas of England include the following stretches of coastline notable for their erosion problems (The Royal Commission on Coastal Erosion, 1911; Steers, 1948; King, 1972):

Fylde Coast. Previous severe erosion (about 1m/yr) of glacial clays. Now extensively protected.

Wirral Peninsular. Erosion of glacial clays and coastal sands now progressively checked by engineering works.

Hampshire Basin. Poorly consolidated Quaternary clays and sands from Poole to Selsey, now subject to extensive protection works.

South Wight Coast. Chalk and underlying Cretaceous sands and clays are undergoing erosion (about 0.4/yr) in unprotected areas.

Brighton to Beachy Head. Chalk cliff eroding, where unprotected, at about 0.5/yr.

East Anglia, Orfordness to Cromer. A cliff-line of poorly consolidated Quaternary sands and clays, receding at rates of metres each year and only protected around coastal development sites.

Lincolnshire Coast. The shoreline, cut into glacial clays and post-glacial clays, peats and sands, is retreating in areas at rates of 1.5m/yr.

Holderness Coast. An elongate coastline of glacial drift cliffs, almost entirely unprotected, and eroding at 1.5-2.0m/yr.

North Yorkshire Coast. The alternate exposure of Jurassic shales and glacial clays is irregularly but steadily eroding at rates of 0.05m/yr.

In all these areas serious erosion is known to have been active for many centuries. Offshore sand and gravel dredging, which has only taken place on any scale in the last thirty years, has however been looked on with suspicion by the inhabitants of these problem shorelines (Figure 9.1). In consequence it is necessary to clarify the position of offshore dredging in relation to adjacent shoreline stability. As a first step, a broad understanding of the dynamics of the nearshore environment is required.

9.2 THE DYNAMICS OF BEACH AND NEARSHORE ENVIRONMENTS

9.2.1 Waves at the shoreline

Offshore wave climate. The characteristics of wind generated waves depend on the strength and duration of the wind, the fetch (the length of open water over which waves are able to develop) and the period of time the wave has travelled onwards after the wind has ceased. The largest waves are formed under very strong winds in deep water where there is a considerable fetch. Waves experienced off the UK coast can be divided simplistically into two categories:

(i) 'Sea'; waves within their area of generation, characterised by their steepness and confused nature. Wave period typically ranges between 3 and 7 seconds.

(ii) 'Swell'; waves may travel many thousands of miles after generation. Wave length remains unaltered but height is lost (giving a long, low profile) and crest patterns become more regular. Wave period commonly lies in the range 7-12 seconds in UK waters.

The effects of friction. Orbital wave motion penetrates deep into the water column. As waves move into shallow water they begin to 'feel' the bottom, which results in three phenomena:

(i) Orbital motion is translated into to-and-fro movement at the seabed, which increases in magnitude as the water shallows (Figure 9.2). At first water motion is purely oscillatory, but as the wave continues to shoal the flow becomes asymmetrical, with a shorter more powerful landward and longer weaker seaward component.

(ii) Wave velocity decreases as a function of the water depths, refracting the wave trains so that crests parallel the shore (Figure 9.3). Refraction may or may not be complete by the time the shoreline is reached.

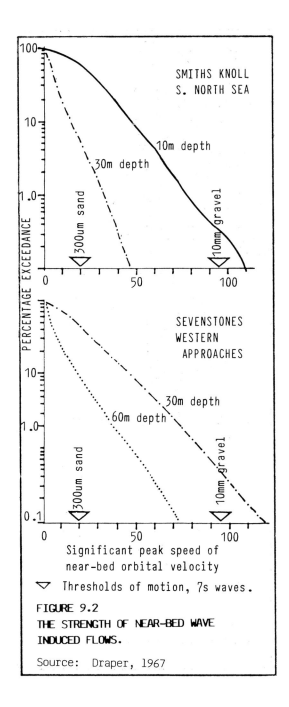

FIGURE 9.2
THE STRENGTH OF NEAR-BED WAVE INDUCED FLOWS.

Source: Draper, 1967

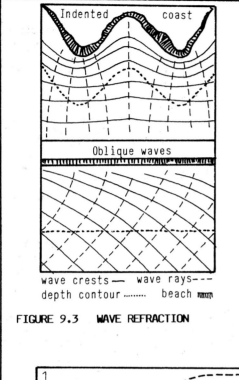

FIGURE 9.3 WAVE REFRACTION

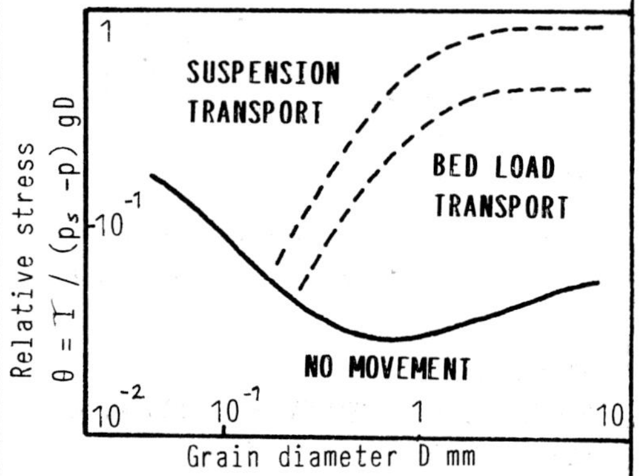

FIGURE 9.4 THRESHOLD AND MODE OF TRANSPORT OF DIFFERENT GRAIN SIZES UNDER VARYING SHEAR STRESS. DERIVED FROM LABORATORY FLUME RESEARCH.
Source: Komar, 1976

(iii) With the complex interaction of changing friction and velocity, the characteristics of individual waves are modified; wave period remains static, wave length steadily decreases and wave height tends to rapidly increase into very shallow water until breaking occurs. Waves break in depths typically 0.7 to 1.4 times their inshore wave height. A train a waves breaking on a shore can set up local water circulation patterns, independent of tidal flow.

These three phenomena control the distribution and magnitude of wave energies arriving at a shoreline, whether a steep rocky cliff-line or a wide sandy beach. In so doing they dictate the erosion potential, and the mechanisms of beach and nearshore sediment transport.

9.2.2   Beach characteristics

A **beach** is a potentially mobile body of intertidal and shallow subtidal sediments, formation of which is attributable primarily to the action of waves against a shoreline.

The **persistence** and **dimensions** of a beach reflect a long-term balance between the rate at which beach forming materials are supplied and removed from the area. A beach supply may be:

(i) inherited - the beach may be a reworked older deposit, or may have been progressively retained during the past thousands of years of the post-glacial (Flandrian) marine transgression,
(ii) derived from local cliff erosion,
(iii) riverborne,
(iv) presently supplied from offshore; the asymmetry of oscillatory flow beneath shoaling waves imparts a residual shoreward motion on coarse (bedload) particles in motion.

Beaches may be depleted by:

(i) intertidal transport along the coast; the breaking of waves at an angle to the shoreline effects this longshore drift, which is only checked at beach-free headlands, groynes etc,
(ii) transport into estuaries opening onto the beach,
(iii) attrition, whereby beach sediments are ground below a critical size (about 180um) and can become dispersed offshore in suspension.

The relative contribution of sources and depletion mechanisms will vary between beaches.

The **particle size** of a beach. The minimum size is approximately 180um, below which wave action effectively winnows particles out to sea. The maximum size varies between about 1mm on a well-sorted beach of long-travelled particles, to huge boulders derived from cliff falls. The particle size characteristics of the beach affect the way it can be shaped by the waves. Sandy beaches are wide and flat whereas shingle beaches are narrow and steep.

The presence of a substantial beach is an important factor controlling coastal erosion, as it can absorb and dissipate storm wave energies and thus prevent retreat of the land.

9.2.3   Beach and nearshore sediment exchange

Flowing water, whether a unidirectional tidal or wind-wave driven current, or oscillatory wave-induced motion, exerts a shearing stress on the seabed. Once the critical shear for a given particle-size is exceeded sediment transport occurs. The rate of transport varies with the excess shear stress, movement initially taking place as slow bedload creep, but with sufficient flow energy water turbulence is sufficient to maintain particles in suspension and effect rapid transport. Particles of less than about 180um diameter pass easily into suspension once the critical shear has been exceeded, but larger particles require progressively greater amounts of energy to leave the bed. These inter-relationships are summarised in Figure 9.4.

Onshore-offshore wave-driven sediment transport. Four shore-parrallel zones can be recognised when a train of waves breaks along a beach. Innermost is the swash zone, which dries and inundates with the arrival of each wave; then the surf zone which extends out as far as the most seaward breaking waves; beyond which extends a transition zone which merges into the nearshore zone (Figure 9.5). Each zone plays a distinct role in terms of sediment transport as follows:

Swash zone. Here energies are intense, and are sufficient to move all sizes of material up and down the beach face.

Surf zone. This zone may be very wide during storms, with waves breaking and reforming several times. The high energy turbulent flows are however insufficient to effectively maintain shingle in suspension, and the residual motion of this sized material is

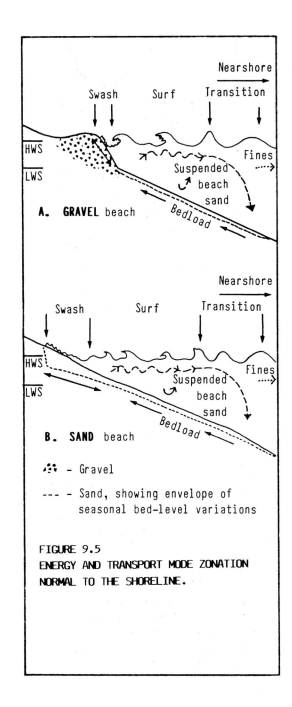

FIGURE 9.5
ENERGY AND TRANSPORT MODE ZONATION NORMAL TO THE SHORELINE.

SITE USA	H̄	T̄	d
Atlantic Coast			
Assateague, MD	0.66	8.73	5.4
Bull Island, SC	0.72	6.17	4.2
Tybee Lighthouse, GA	0.85	7.30	7.4
Boca Raton, FL	0.48	5.44	4.2
Gulf of Mexico Coast			
St. Andrews Park, FL	0.53	4.86	4.5
Crystal Beach, FL	0.52	4.81	4.8
Gilchrist, TX	0.39	6.86	4.1
Galveston, TX	0.47	6.71	3.8
Corpus Christi, TX	0.79	6.66	5.2
Pacific Coast			
San Clemente, CA	0.87	15.16	7.1
Bolsa Chica, CA	0.74	12.10	6.1
Pt. Mugu, CA	0.86	14.69	5.7
Pismo Beach, CA	0.93	12.03	7.5
San Simeon, CA	0.94	11.58	6.5
Capitola Beach, CA	0.46	11.16	4.2
Stinson Beach, CA	1.15	12.62	7.3
Wright's Beach, CA	1.50	11.37	10.4
Shelter Cove, CA	0.71	11.88	7.1
Prairie Creek, CA	1.05	10.69	7.0
Umpqua, OR	1.14	9.38	7.8

TABLE 9.1 CALCULATION OF THE SEAWARD LIMIT OF BEACH AND SAND EXCHANGE FROM WAVE STATISTICS.

$\bar{H}$: Mean wave height (m).
$\bar{T}$: Mean wave period (sec.)
d: Water depth to which beach sand may annually migrate (m).

Source: Hallermeier, 1981

landwards. Sand, on the contrary, can be held in suspension and efficiently dispersed out to the seaward limits of the breakers. This occurs particularly during periods of large, steep storm waves, which tend to erode the upper beach face and comb material seawards. Long, low fair weather swells reverse this process and carry sand back onto the beach.

Transition zone. The turbulence derived from wave breaking dies out across this zone in a seaward direction. Medium sand carried through the breakers during storms therefore resettles to the bed and is again subjected to the residual landward bedload movement due to the asymmetry of the wave orbital motion. Only material finer than about 180um, which is more readily kept in suspension, continues to be transported off or alongshore in the water column.

Nearshore zone. This is commonly the area of fallout for fine sand. Orbital wave velocities may only be sufficiently strong to initiate seabed motion during the stormier periods of the year, but are however sufficient to prevent the long-term accumulation of mud.

To summarise, shingle moves as bedload only and with a landward residual in all zones except the swash zone (Figure 9.5A). Sand (>180um) however may move as bedload or suspended load in the swash and surf zone, tending to move seawards during winter and inshore during summer. Suspended load transport of this material ceases across the transition zone, defining a seaward limit to the annual migration of beach sand. This sand can move as bedload with a landward residual in both the transition and nearshore zones (sand movements are shown in Figure 9.5B). Fine sand and mud are effectively winnowed out of the swash, surf and transition zones, and accumulate in the nearshore zone or further afield. The widths and efficiency of transport within each zone increase with wave height.

Alongshore transport. **Littoral drift** effects the movement of sand and shingle along the beach within the swash zone. **Wave-drive littoral circulation systems** can produce shore-parallel transport of sand within the breaker and transition zones. **Tidal currents** can transport sand, and under certain circumstances gravel (Hammond et al, 1984), parallel to the coastline in the transiton and nearshore zones (shallowness of the water and wave-interference normally reducing the efficiency of tidal flow close inshore).

The effective **seaward extent of seasonal beach sand migration** (the landward edge of the nearshore zone) is an important parameter. It may often be recognised in the field from particle size characteristics of local sediments, but may be masked in areas of strong tidal infuence. Hallermeier (1981) has derived a formula based on recorded wave statistics which appears to realistically predict this limit. Examples from various shore locations in the USA are shown in Tabe 9.1; beach sand exchange is typically confined within the 3-10m isobath.

9.2.4 The supply of beach sediment from offshore

In the preceding sections it has become clear that wave action in the nearshore zone is capable of transmitting beach-forming sediments shorewards. The significance and rates of this potential beach supply are poorly understood for several reasons;

(i) Although it is possbile to apply known sediment movement thresholds to predicted oscillatory wave motion at depth, this only enables prediction of wave **disturbance** of sediments, and not landward residual transport. For example, from Figure 9.2, at 30m depth sand will be disturbed for about 5% of the year in the southern North Sea and 50% of the year in the Western Approaches, but the direction of any wave-induced direction of movement is unknown. For 10mm gravel, the respective disturbance times at 30m depth are 0% and 0.5%; and at 10m depth 0.3% and >1%, with direction and rate of movement again being unknown.

(ii) Direction and rate of onshore sediment transport within the wave-disturbed zone will result from **the combined effects of wave and tidal currents.** The study of this complex interaction is in its infancy. Laboratory investigations have as yet only examined the effects of superimposition of aligned wave and tidal flows (B A O'Connor pers. comm., Collins and Hammond,1979), whereas in shoaling waters the two flows will tend to lie normal to one another. Prediction of transport is not yet feasible.

(iii) **The role of the extensive fine sand areas,** that invariably lie seaward of beaches, in absorbing energies and inhibiting the bed load transport of coarse material has not been evaluated.

Although many beaches are known to contain a high proportion of offshore-derived sediments, the relative importance of present and past mechanisms and rates of supply (for sands in particular) are largely unknown.

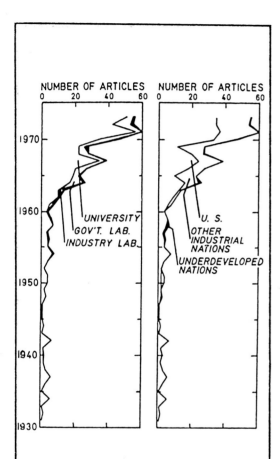

FIGURE 9.6 NUMBER OF ARTICLES ON CONTINENTAL SHELF SEDIMENTOLOGY PUBLISHED PER YEAR IN SELECTED JOURNALS, AS AN INDEX OF RESEARCH EFFORT.

Source: Emery, 1976

9.2.5   Man-induced coastal change

Since man has developed the ability to undertake coastal civil engineering works on a large scale he has affected nearshore dynamics and coastal stability. Commonly cited examples are sea walls, which check cliff erosion and stabilise a locality, but to the detriment of down-drift areas whose supply of beach material has been cut off leading to beach depletion and erosion. Similarly groynes will slow down littoral drift in their vicinity, building up beach levels and reducing erosion, but at the expense of down-drift beach stability. The alteration of the nearshore energy climate by the construction of islands, training walls and dredged channels also has repercussions for the stability of adjacent coasts (O'Connor,1985), which may be positive or negative.

Negative effects attributable to aggregate dredging are rare. The classic case is Hallsands, where over a period of some thirty years following 1872, $3 \times 10^3 m^3$ of shingle were removed from the lower (intertidal) beach face during the construction of Devonport dockyard. The consequential reduction in beach levels resulted in the sea destroying the adjacent fishing village. A more recent example is cited from Botany Bay in Australia, where changed wave refraction patterns due to inshore dredging were shown to be responsible for coast erosion (Brampton,1985). Because of this potential effect, all dredging licence applications for UK shelf waters are examined in relation to the stability of the adjoining coast (Section 1.3.1).

9.3   DREDGING RELATED RESEARCH

9.3.1   The history of research

The intensity of research into continental shelf processes generally has increased markedly since about 1960, both nationally and internationally (Figure 10.6). In Great Britain research specifically related to aggregate dredging and shoreline stability has been almost exclusively undertaken by Hydraulics Research Ltd (HR, previously the Hydraulics Research Station under the DoE). The Institute of Oceanographic Sciences (IOS) and several university departments have however undertaken research in related subjects, such as offshore gravel mobility. Research undertaken by HR has been commissioned by the Minerals Division of the DoE, and the Crown Estate Commissioners (CEC), the latter usually in relation to specific licence proposals and funded by the companies involved.

The first research on the effects of aggregate extraction was undertaken by HR in 1968. Licences had of course been issued before that date. Some of these have been subsequently reviewed by HR (eg Solent, Pot and Prince Consort Bank, HR,1977) and others allowed to continue on a 'no reported adverse effect' basis. All licence applications since 1968 have been considered by HR, which now possesses a reasonable database on offshore dredging and related coastline effects.

In 1976 a report by the Verney Committee on Aggregates (DoE,1986) recommended:

(i) "the existing studies of the Hydraulic Research Station should be expanded with a view to establishing in the light of growing knowledge whether the large areas sterilised for coast protection reasons - particularly inshore areas - may now be dredged without the risk of unacceptable damage to the coast. To this end all existing refusals on coast protection grounds should be reviewed and classified as:

(a) proven and
(b) not proven.

Each area designated not proven should then be reconsidered and experiments carried out to prove one way or the other whether the fears are justified; and

(ii) the possibility that a programme of beach nourishment might lead to the release of current sterilised reserves should be examined.

It is important that a programme on these lines should be pursued urgently and with vigour and not allowed to falter through lack of funds. The Crown Estates Commissioners, as good landlords, should be invited to contribute to this programme."

The response from the Secretary of State made it clear that such a programme of work was not justified in the then current economic circumstances; the same viewpoint has remained prevalent.

9.3.2   Research results

Field investigations have been directed towards determining the seaward limit of the active supply of sediment to beaches (Sections 9.2.3, 9.2.4).

(i) Tracing of radioactively labelled shingle; off Worthing 1969 (HR,1972). Shingle (25mm) was placed at 9, 12, 15 and 18m depths in an area of weak tidal action, and its movement tracked over the subsequent 20 months. Wave observations from the

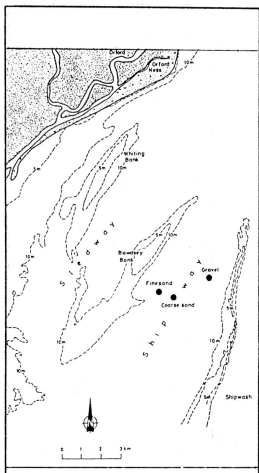

- Radioactively labelled sediment release sites

FIGURE 9.7 THE SHIPWAY RESEARCH AREA

Source: HR, 1985

Owers Light Vessel were used to relate the rate of shingle movement to prevailing wave conditions. The results demonstrated an increase in shingle mobility with decreasing water depth. At 18m depth there was no movement of gravel. At the 9 and 12m sites a small net landward residual was detected in the order of 1km per 60 years. The study concluded that on the south coast the movement of shingle in water deeper than 18m will be negligible at all times.

(ii) Tracing of radioactively labelled shingle, coarse sand and fine sand; Shipway Channel, 1982 (HR, 1985). Tagged material was laid at three sites between the Shipwash and Bawdsey Banks (Figure 9.7), in a water depth of 13m, and movements traced for one year. Peak near-surface tidal streams (from Admiralty data) are 90cm/s (flood) and 110cm/s (ebb). Wave conditions were predicted from wind records at Shoeburyness. There was no evidence that the shingle (38-19mm) or coarse sand (400-1000um) underwent any movement, but the fine sand did diffuse along the axis of the tidal flow. The shingle stability is not surprising, confirming earlier work in the area (Kidson and Carr, 1959). The lack of sand transport is unusual considering the strength of tidal flow, and is attributable to the shielding effect of the natural gravel substrate, the small quantities of tracer used becoming trapped within the gravel interstices. The results demonstrated the viability of dredging in water shallower than 18m in relatively sheltered outer estuary areas.

(iii) Current-meter measurements, south of the Isle of Wight (Jaffrey, Motyka and Price, 1978). A series of measurements were made in water depths ranging from 16 to 29m, with a recording current-meter at 0.4 x depth. Wave data was obtained from the nearby Owers Light Vessel. A theoretical approach was used to calculate the shear stress at the seabed due to the combined action of tides and waves. The preliminary results suggest that under south coast wave conditions and surface tidal velocities in the order of 100-150cm/s 25mm shingle may be mobile down to 22m water depth.

Physical and numerical modelling has been directed towards identifying the effects on adjacent coast of changed wave refraction due to dredging (Section 9.2.1).

(i) Numerical modelling of wave refraction. HR have developed a standard wave refraction computer programme which plots the changing direction of movement and height of waves as they move into shallow waters. The input to this model is data on the offshore wave climate and data on the seabed bathymetry in a grid form. Where the former is not directly measured, techniques of increasing sophistication are used to predict relevant characteristics of the wave climate from Meteorolgical Office wind records (Seymour, 1977; Hasselman et al, 1973). Wave rays can be back-tracked from known sensitive stretches of coastline and the normal change in wave height and direction in those rays passing over the proposed dredging area recorded. The seabed topography data input to the computer is then modified according to the dredging proposals, and the resulting changes in wave height and direction of approach to the coastline identified. In this way dredging proposals which have no measurable effect on the wave climate of adjacent coasts can be identified (HR,1984). The model, because of contained simplification, is known to exaggerate refraction effects, thus building a safety factor into prediction.

(ii) Numerical modelling of shoreline changes due to wave refraction. In an attempt to identify potential changes in beach plan resulting from changed wave refraction over dredged holes, the above wave refraction model was coupled with a beach mathematical model in a hypothetical environment (Motyka and Willis,1974; HR,1976). A wave climate typical of the southern North Sea, an onshore-offshore profile approximating that between Great Yarmouth and Cross Sands, and an initially straight plan beach were imput to the model, and the effects on beach plan of dredging holes of various depths and geometries at various positions offshore were examined. For the set of conditions tested, the effects of dredging in water depths greater than 14m were negligible irrespective of the depth and extent of excavation. In shallower water a longshore pattern of alternating beach accretion and erosion was generated, with net erosion increasing with proximity of dredging to the shoreline (Figure 9.8). Close to the beach, erosion was found to increase with length of hole parallel to the shore and with depth of hole. Because of the simplifications employed, the results of this research are used only in an illustrative fashion, and the 14m criterion is not applied to actual situations. Similar work has been undertaken in Japan (Horikawa et al,1981).

(iii) Numerical and physical modelling of wave attenuation in areas of offshore banks. This project (HR,1983) was undertaken with the aim of providing a better understanding of the way in which offshore banks protect adjacent shorelines from wave action, and to investigate the effects of dredging a channel area between the banks. A physical model was initially used to determine the relative importance of the various ways in which wave energy was dissipated by the bank; friction effects were demonstrated to be most important, and also wave breaking at shallowest modelled depths over the bank crest. Wave reflection was shown to be surprisingly unimportant. The HR wave refraction computer model was then used to simulate conditions over a wide range of wave heights, periods and approach angles, the forward-plotting ray model used allowing incorporation of friction and breaking effects. The model demonstrates convincingly the important role of the Shipwash Bank in reducing wave heights in its lee (Figure 9.9), emphasising the importance of avoiding bank crests when dredging. The modelling also showed that dredging to a depth of one and perhaps two metres across an extensive channel floor area in the lee of the bank would have no effect on waves arriving at the adjoining coast.

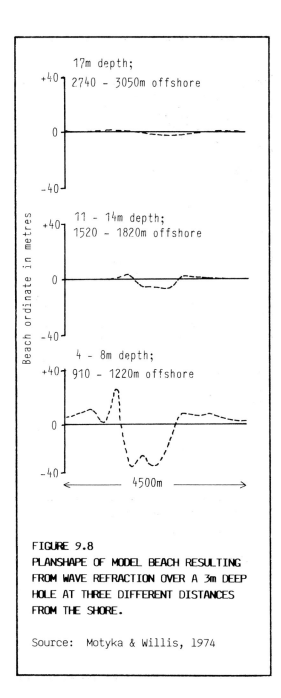

FIGURE 9.8
PLANSHAPE OF MODEL BEACH RESULTING FROM WAVE REFRACTION OVER A 3m DEEP HOLE AT THREE DIFFERENT DISTANCES FROM THE SHORE.

Source: Motyka & Willis, 1974

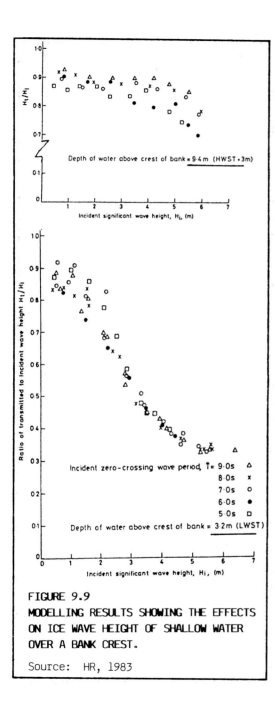

FIGURE 9.9
MODELLING RESULTS SHOWING THE EFFECTS ON ICE WAVE HEIGHT OF SHALLOW WATER OVER A BANK CREST.

Source: HR, 1983

9.3.3   Research and licensing

Procedure. A licence application for dredging (Section 1.3.1) is passed to HR at an early stage by CEC through their technical adviser for an opinion as to whether dredging is likely to affect the adjacent coastline.

Following an initial assessment by HR (which often includes informal consultation with the local body responsible for shoreline protection) the extent of research necessary is agreed with CEC advisers and may include a desk study, computer simulation, site inspection or a full-scale field research programme. This information is passed by CEC to the dredging company who decide whether the cost of the research is ecomonically justified.

HR assesses the likely impact of the proposal (defined by an area of extraction and a maximum extraction rate) on the basis of four criteria (Bramptom,1985);

(i) Is the area within the zone of seasonal beach sediment exchange (in or shoreward of the Transition Zone, Section 9.2.3)? If yes, beach drawdown is likely to occur, resulting from the non-return in summer of material combed off the beach during the winter season, and the application is regarded unfavourably.

(ii) Is shingle mobile at the dredge site? This is assessed in the light of known tidal currents and wave climate (Section 9.2.4). In the absence of specific measurements of shingle dispersion, and on open coasts as opposed to areas of offshore banks, the 18m depth criterion is usually relied upon. If the answer is yes, a potential supply of shingle to the shore is considered to be at risk and the proposal is regarded unfavourably.

(iii) Does the dredging area include banks which if removed would increase wave activity at the shoreline? If the answer is yes then the proposal is usually advised against, unless unusual mitigating circumstances are evident.

(iv) Is the area sufficiently distant from the shore and in deep enough water so that changes in wave refraction (Section 9.2.1) over it do not lead to changes in beach plan shape? To check this it is usually necessary to run the standard HR wave refraction computer model for the wave and seabed conditions of the proposed site. If the proposed bed alterations result in measurable changes in the distribution of energy at the shoreline then the application is regarded unfavourably.

The views of HR (which may include various qualifications) are considered by CEC. Where it is appropriate, informal discussions are held between CEC advisers, HR and dredging company to consider any qualifications or possible modifications to the proposed licence which would enable the CEC to accept the application for the 'Government View' procedure (Section 1.3.1).

The **Zero Effect Principle.** It is readily apparent that the decission communicated to the applicant dredging company as a result of the above assessment is intended to err well on the side of caution. This is found in both the application of science, where modelling is known to overestimate effects, and from the requirement that no measurable change in the wave climate or the sediment regime of the coast must result from the dredging operation. In the Consultants' opinion the application of this zero effect principle probably prevents aggregate extraction in many areas where it would in reality be safe to work in terms of the stability of the local shoreline. Beyond this, it may be suggested by mineral planning authorities, under pressure from both aggregate companies and environmental pressure groups on land, that allowing minimal, controlled effects along a low-value shoreline may be more acceptable than the opening of new agricultural areas for extraction (the 'balance sheet' approach, Section 13.4.1). **The present state of development of coastal sedimentology, oceanography and engineering, and its foreseeable progress in the next fifteen years, does not and will not allow such accurate prediction and control to be made.**

> "The complexity of the coastal environment means that computer models are likely to evolve slowly, with a continuous updating every few years as engineers improve their understanding of the environment. It should also be remembered that research into sediment movements in rivers has been in progress for more than 100 years in many institutions throughout the world and yet there is still no universal theory available, which is capable of describing transport rates to an accuracy of better than a factor of two." (O'Connor, 1985).

Against this background, it is only reasonable that the no-effect criterion remains in force.

The effects of insufficient data and conflicting scientific opinion. When informed of a licence refusal or of the necessity to impose restrictions on operations, an applicant is told whether this is due to lack of data on the environment. If so three courses are then open to the dredging company.

(i) To wait over a period of years and see whether information generated elsewhere will help the case.

Marine aggregate is frequently used in beach replenishment schemes

(ii) To commission the necessary fieldwork. Whereas initial HR consultancy, including the use of computer modelling, may be achieved for less then £5000, involvement in large-scale field investigations could cost upwards of £50,000. No field investigations have yet been commissioned by the Industry.

(iii) To query HR's conclusion on the basis of the existing data.

In relation to the last point, it is also possible for an objector to the granting of a dredging licence to seek alternative expert views to check the 'safety' conclusions arrived at by HR. **Instances of both have occurred in the past two years, apparently reflecting an increasing awareness that where environmental processes are concerned 'a fact' is a rare phenonemon, and that most decisions are based upon probalility predictions.** The latter are best arrived at from an amalgamation of views, derived from the consideration of problems from various angles, thus this new trend is, in the Consultants' opinion, a healthy one.

9.4   SHORELINE CONSIDERATIONS IN FUTURE PERSPECTIVE

The estimated total tonnage of aggregate removed from the UK Continental Shelf since offshore dredging began is equivalent to a box 1m deep and 15km x 15km in extent cut into the seabed. At present rates this 1m deep box is expanding by about 9km² each year. Seen in relation to the total length of coastline (4,300km in England and Wales) the potential for damage is seen to be small - **concern that exists is on a local not a national scale.**

Even in dredging areas that were established prior to 1968, and which probably would not pass the present day 'safety' criteria, no coastal erosion has yet been scientifically attributed to the effects of dredging. **There is no complacency regarding the efficiency of the administered 'safety margin' however, and at least two areas of concern have been identified.**

(i) The need to define the geometry of mined holes in the seabed. The depth and shape of extraction areas can affect wave refraction patterns, particularly in the shallower waters worked by dredgers (Section 9.3.2). When asked to undertake wave refraction models, HR are forced through insufficient data to divide the proposed production tonnage by the licensed area to determine the average depth of working. Clearly this is not ideal, and in some circumstances localised concentraton of mining effects could change the outcome of the refraction tests. The remedy is the generation of detailed prospecting data and the definition of more precise mining proposals.

(ii) The combined role of tidal and wave-induced currents in transporting sediments, particularly in relation to sands. The assumption that tidal currents run parallel to the coast and therefore play a minor role in the supply of beach material has rather simplistically been relied upon to date in the virtual elimination of such considerations from the HR assessment. Budgets of sand in motion along a coast, potentially capable of supplying beaches (Section 9.2.4), are generally poorly understood (Heathershaw et al, 1981; Lees, 1983), and the conversion of a 'sand in transit' zone to a 'sand trap' zone as may occur with dredging is, in priciple, undesirable. The generaton of local data during prospecting (side-scan surveys, current-meter recordings, near-bed suspended sand measurements) would be of tremendous help in assessing the importance of such pathways within proposed dredging areas.

From the dredging operators point of view, large areas of resources are currently unattainable because of coastal stability objections (six licence applications have been turned down to date for this reason). **Most rejected applications will have been turned down because of lack of suitable data** rather than a conviction on the part of HR that coastal erosion would have resulted. Within the constraints recognised in the previous section, the way forward lies logically with increased data generation and assessment. This may be achieved in two ways (not necessarily mutually exclusive).

(i) **Site-specific studies can be pursued, funded by individual dredging companies.** For the future such work can be most economically realised during the latter part of a prospecting programme. Sufficient reliable expertise is now available within commercial settings in the UK for this type of work to be undertaken on a competitive tender basis, thus optimising cost efficiency.

(ii) **Fundamental research could usefully be undertaken:** topics suggested include the determination of a numerical relationship between gravel mobility and bathymetry, wave climate and tidal regime, in order to relieve dependence on the not universally applicable '18m' concept (necessarily including evaluation of the combined effects of wave and tide action on sediment mobility); investigations of the effects of hole geometry on refraction patterns; the role of offshore supply in the sand budgets of sea beaches and inshore banks.

Research priorities can only be firmly established with reference to the details of past licence refusals; this can either be undertaken in confidence at HR, or within the context of individual company experience.

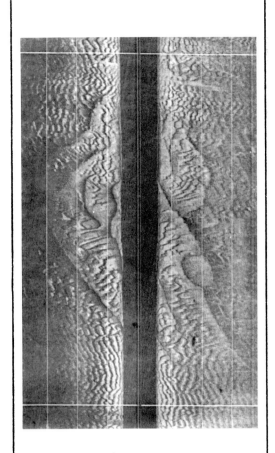

Sand transport on the seabed shown by side-scan sonar data

In view of the high costs of field investigations, and the constant generation of new ideas for field measurement and date processing techniques, **a review of methods currently available for aggregate extraction related research** could promote more cost efficient studies. As an example, the use offshore of aluminium pebbles and remote recording metal detecting equipment (Wright, Cross and Webber,1978) may provide a less expensive means of establishing shingle stability on the seabed than currently employed radioactive tracers.

FISHERIES AND ENVIRONMENTAL QUALITY

10.1   THE ORGANISATION OF PROTECTION AND RESEARCH

10.1.1   International

"The exploration of the continental shelf and the exploitation of its natural resources must not result in any unjustifiable interference with navigation, fishing or the conservation of the living resources of the sea..." Article 5.1, Geneva Convention on the Continental Shelf, 1958.

Beyond this statement of general principle, no common legislation exists for controlling marine mining in different sectors of the European Shelf, in contrast to areas such as waste disposal and fishery exploitation. Control over the latter activities has been introduced during the last 10-20 years, and it is possible that pressures resulting from increased marine aggregate production could lead to similar European legislation in the foreseeable future. It is therefore important to bear this consideration in mind when planning for the present day national situation.

In the 1970s the **International Council for the Exploration of the Sea** formed a Working Party in an attempt to draw together an international fisheries view regarding marine aggregate extraction (ICES 1975, 1977, 1979, De Groot 1981). After a useful start the group was allowed to lapse, presumably as a result of the economic recession in Europe and decreased pressures for new dredging areas. In its various reports the Working Party has highlighted national research efforts into the effects of offshore aggregate extraction. The Working Party was however criticised for its narrow basis, and many of its initial conclusions were strongly refuted by the British Industry (Anon, 1979). In May 1986 the Working Party was reconvened, and now has a much broader membership.

10.1.2   National

The control of dredging in relation to United Kingdom sea fisheries and conservation interests is effected through consultation with the Ministry of Agriculture, Fisheries and Food (MAFF, for England and Wales) and the Department of Agriculture and Fisheries for Scotland (DAFS) during the 'Government View' procedure (Section 1.3.1). It should be noted that these bodies have the specific brief

---

10

CONTENTS:

The organisation of protection and research
  International
  National

Dredging and general marine environmental health
  Definitions
  Effects on water quality
  Destruction of habitat
  Recolonisation of the seabed

Dredging and fish stocks
  Shellfish
  Adult fish
  Young fish
  Spawning

Potential conflict with the inshore fishing industry
  Description of the inshore fishery.
  Operational conflict
  Long-term sterilisation of bottom fishing grounds.

The outlook for the future

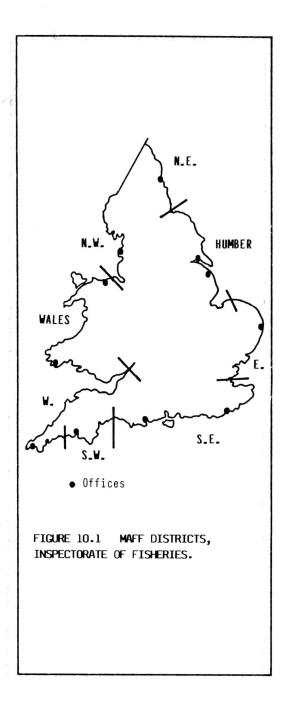

FIGURE 10.1   MAFF DISTRICTS, INSPECTORATE OF FISHERIES.

"To protect the marine environment and biota, including the fish stocks, against pollution; to undertake adequate monitoring in order to maintain a check on the state of the marine environment; and to minimise interference with fishing" (MAFF, 1985).

MAFF/DAFS activities cover the wide range of interactions between the fishing industry, conservation interests and other users of the sea. Problems faced relate to fish stock management, waste disposal at sea (including radio-active discharges) as well as dredging related issues.

Whereas in broad terms fishery management and environmental conservation go hand in glove, in specific contexts there may be areas of divergence where conservation issues are not adequately covered. This has been pointed out by the Nature Conservancy Council (NCC, 1979), which body may play an increased role in licensing control in future years.

MAFF (DAFS in Scotland). The administrative headquarters of these government departments (in London and Edinburgh respectively) co-ordinate the fisheries response to dredging licence applications (through the DoE, Section 1.3.1), and also handle fisheries issues arising from dredging operations (dealing with the Crown Estate Commissioners). The MAFF headquarters interfaces with the fishing industry via a system of District Inspectors (DIs - Figure 10.1), and a similar system applies in Scotland. Scientific advice is derived from the staff of a series of research laboratories; those at Lowestoft and Burnham-on-Crouch have been involved in marine aggregate programmes.

Research into fisheries aspects of sand and gravel mining at sea was initiated in the early 1970s, in response to growing disquiet as to the effects of dredging. Work was organised under the Aquatic Environment Commission objective of providing scientific information and advice on non-polluting man-made changes in the environment. **The budget for the programme remained small, in financial terms never exceeding 1% of all commissioned research.** Work had ceased by 1980. Several useful projects were completed during the decade, including:

- observation of stability of isolated dredged pits and furrows; from sector scanner observations and measurement of tidally induced shear stresses (Dickson and Lee, 1973).

- measurement of water quality in a dredging area (Milner, Dickson and Rolfe, 1977).

- comparison of benthos and fish feeding patterns between control and dredged areas (Shelton and Rolfe, 1972; Milner, Dickson and Rolfe, 1977; ICES, 1979).
- laboratory study of sandeel spawning (R Milner, MAFF, pers.comm.).

NCC. In 1979 the joint NCC/National Environmental Research Council (NERC) Working Party on Marine Wildlife Conservation reported its findings (NCC, 1979). Three strategies for marine conservation were put forward:

(i) The establishment of Marine Nature Reserves (MNRs).
(ii) The protection of particular species.
(iii) The promotion of responsible use of natural resources.

The aims of (i) were fulfilled with the passing of the Wildlife and Countryside Act in 1981, which made it possible for the first time to establish statutory Marine Nature Reserves over areas

"covered (continuously or intermittently) by tidal waters or parts of the sea in or adjacent to Great Britain up to the seaward limits of territorial waters".

To achieve conservation aims Marine Nature Reserves in an offshore situation will necessarily be large (many square nautical miles in extent). The effect from the aggregate dredging point of view may be to bar areas of the continental shelf as potential working sites. The extent and number of MNRs that will ultimately be established are unknown. Equally, the priorities relied upon by the Secretaries of State in the decision making process are untested. The NCC has already expressed interest in several sea areas, including Lundy and the St.Abbs Head area, both possibly containing viable aggregate resources. The NCC are also consulted regarding dredging within estuaries (DoE, 1986) and wharf siting issues (Section 11.4.3) as part of the present 'government view' procedure.

NERC. The Council distributes government research funds in the field of the natural sciences, supporting organisations such as the Institute of Oceanographic Sciences (IOS), the Institute for Marine Environmental Research (IMER), Universities and an equipment/research vessel pool. Traditionally concentrating on pure research, the Council in line with present-day thinking now encourages more practical programmes. Past research relates only in a background sense to dredging questions, but there may be scope for more directly applicable research in the future.

Marine life caught in the spillway gratings

10.2 DREDGING AND GENERAL MARINE ENVIRONMENTAL HEALTH

10.2.1 Definitions and context

Health can be defined on a long and short-term basis. Most biological systems can survive bouts of disturbance with no residual effects, with the possible exception of rare fragile communities inhabiting sheltered niches. A full assessment of the implications of dredging must necessarily consider water column quality, biological characteristics and pelagic fish stocks; seabed biology, demersal fish and shellfish stocks; and particularly the complex interaction between physical system and ecological system changes. In the Consultants' experience three sets of criteria principly dictate the potential duration of bouts of disturbance resulting from dredging:

(i) **What is the extent, duration and intensity of water quality impairment during the dredging period?**
(ii) **To what degree is the physical and chemical environment at and just below the seabed/water interface the same before and after dredging?**
(iii) **At what rate does recolonisation of the benthos take place?**

In the following section these three questions are looked at in the light of the known impact of dredging, with the aim of assessing whether long-term effects occur. Implications for commercial species are then looked at in Section 10.3.

It should be stressed that any potential effects that are identified in the following sections may only be of significance in localised regions. Whereas the area under licence for aggregates extraction is large in relation to coastal fisheries, the area of seabed affected each year by aggregate dredging is only about $9km^2$ (assuming an average depth of removal of 1m - see Section 9.4), which is tiny in comparison to the total UK inshore shelf areas. Thus the scale of dredging impact on the seabed is small, particularly in relation to general environmental health as opposed to fishing. Localised concentration of dredging areas is however a problem in certain areas (Section 10.4.2).

10.2.2 Effects on water quality

Although some degree of impairment of water quality would not be expected to persist significantly beyond the period of dredging (the latter varying from tens of hours over a specific point on the seabed to tens of years, ie the life of a dredging project, for the dredging ground as a whole), the intensity of local water column effects is worthy of consideration, if only as one of the factors controlling the efficiency of post-dredging recolonisation of areas. It is also relevant from the point of view of assessing effects on individual, (particularly commercial) species (Section 10.3).

Turbidity. Sediment rejected during dredging operations consists of sand, which will fall rapidly to the seabed, and silts and clays bound in organomineralogic aggregates (Drake, 1976) capable of remaining in suspension over longer periods. The physical impact of the latter material will depend upon the amount of fines generated, local hydrographic conditions and turbidity regimes, and the susceptibility of the local fauna.

The potential physical effect of high turbidity on life functions (such as feeding, respiration and reproduction) of both seabed and water column dwelling organisms is complex; an excellent review can be found in Moore (1977). It is sufficient to say that animal populations, particularly sedentary and non-migratory species, will be conditioned to their local natural turbidity regime. Any alteration of the latter will induce stress leading to changes in species viability and population structures. Concern must rest with the extent to which the benthos is changed as a result of increased turbidity (affecting the rate at which natural conditions may ultimately be re-established), effects on unique or rare species (conservation issues) and effects on commercial species (either short-term feeding or avoidance problems, affecting adjacent fishing areas, or longer-term disruption of local bottom-dwelling populations - Section 10.3).

Natural turbidity levels vary considerably around the British Isles, and may also fluctuate markedly at a particular site with changing wave and tide conditions. Four broad offshore turbidity regimes may be recognised:

(i) Clear water (in the order of 5mg/l), muddy seabed. Low energy environments (deep water, slow to moderate tides with low diffusion rates) remote from mud sources eg western Lyme Bay.
(ii) Clear water, coarse seabed. High local energies (wave affected, strong tides with associated high diffusion rates) but remote from mud sources eg south of Wight.
(iii) Turbid water (in the order of 100mg/l), coarse seabed. High energy environments as (ii), but close to mud sources eg outer Thames Estuary.
(iv) Turbid water, muddy seabed. Moderate to low energy environments as (i), but dominated by local mud sources eg Channel floors of the inner Thames Estuary.

Regimes (i) and (iv) are not currently worked for aggregates in the UK because of the presence of mud overburden. Should this situation ever change (Section 5.2.2) areas of type (i) may be highly susceptible to the effects of high turbidity discharges.

Clear water overflow

Turbid water overflow

Long-term suspended solids sampler

Areas of type (iii), typified by the outer Thames Estuary, are worked for sand and gravel. Natural suspended solids concentrations have been measured in the Southwold dredging area (Milner, Dickson and Rolfe, 1977) and shown to vary over the tidal cycle between 100 and 600mg/l. These authors concluded that the natural sediment load maintained in suspension by the tides was sufficiently high that outwash material from dredging will not have a significant effect on the environment.

The remaining environment (ii), generalised as being remote from mud sources but of high local energies, as exemplified by the offshore English Channel dredging areas, would appear to pose a problem if a muddy subsurface aggregate deposit was being worked. Even so, the high diffusion rates associated with these grounds could theoretically reduce plume concentrations to ambient levels in an hour or so. No direct observations of dredging plumes have been made from these areas, although such measurements have been made in the Baie de Seine on the French side of the Channel (Augris and Cressard, 1984). Here an experimental area was dredged, with the original surface sediment being medium sand. The generally slacker tidal currents (diffusion coefficients $0.01-0.4m^2/s$)...

> "...mean that the plume maintains its form and disperses only slowly, Thus, around an industrial dredging zone a mass of turbid water persists which moves according to the currents and which only slowly recovers its clarity."

The latter research would appear to confirm that under some restricted circumstances dredging, particularly where several ships may be working in a close area, may significantly affect the local turbidity regime. **These circumstances could be readily identified at the outset of any dredging project by putting together information on the nature of the aggregate reserves with the simplest measurements of the local hydrography and natural turbidity regime.**

Water chemistry. Release of toxins. Fine sediments act as a sink for may pollutants introduced into the marine environment, which become chemically stable within their environment of accumulation and thus effectively removed from the biosphere. Concern has been expressed that during dredging, disturbance and resuspension of muds within the water column will change the chemical environment and lead to the reintroduction of toxins to the sea (De Groot, 1976). However, for conditions found at UK dredging sites there is sufficient awareness to confidently rule out the possibility of toxicity problems arising from present mining practices (M Parker, MAFF, pers.comm.).

Greatest effects related to release of toxins would be expected in mining contaminated seabed areas. The receiving area for sewage sludge and industrial wastes dumped off the mouth of the River Humber (Murray et al, 1979) is partially coincident with the licensed gravel extraction areas. This apparent anomaly is acceptable because of the highly dispersive nature of the Humber dumping area, and would not occur generally. Moreover the Industry does not accept that any area which contained other than minimal amounts of these materials could produce marketable aggregate.

The creation of anoxic conditions. The organic content of buried fine sediments has a high oxygen demand, which is satisfied from sea water oxygen content when the muds are suspended in the water column. Under some conditions (near static water in deep pits in muddy environments eg the Oresund, Ackefors and Fonselius, 1969) this could lead to sites of intense local stagnation, with waters high in dissolved hydrogen sulphide and low in oxygen. Adequate mixing conditions ensure that present British dredging operations could never generate such effects.

Water chemistry problems are therefore not likely to be contentious at present or foreseeable levels of production and using present plant. Should deep extraction from below mud overburden ever be considered then the Consultants feel that the situation would need to be reviewed.

10.2.3 Destruction of habitat

There are three possible effects of aggregate dredging on the habitat potential of extraction and adjacent areas:

(i) Changes occur due to the removal of the existing substrate and the uncovering of underlying strata.
(ii) Changes occur due to the burial of the existing substrate with material mined and rejected overboard.
(iii) There is no change, the physical characteristics of the substrate remaining essentially similar before and after the extraction.

In an ideal world, where unknowns are kept to a minimum, the 'no change' criterion would be enforced. This has not happened in the past, and as will be seen cannot be relied upon in the future. Thus, in determining the acceptability of a dredging proposal, the environmental value of the new habitat in relation to the old must often be assessed.

Grab sampling during MAFF benthos survey

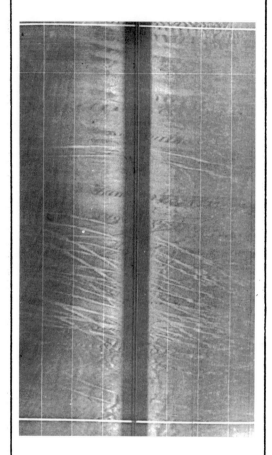

Side-scan sonar record of a lightly trailer dredged area, showing localised concentration of effort

Vertical scale; 1mm to 1.5m
Horizontal scale; 1mm to 5m

Removal of the existing substrate and the uncovering of underlying strata. The most obvious way in which substrate change may take place results from localised total removal of the gravel deposit, resulting in the exposure of underlying material. In a worst case the underlying strata may be a barren stiff clay, rock or mobile sand, with poor recolonisation potential. More subtle changes in the relative percentages of gravel, sand and mud would cause less drastic faunal changes, possibly advantageous, possibly not.

In the past the dredging of anchor pits has been known to expose underlying muds (Shelton and Rolfe, 1972); in practice this would not occur very often as at the first signs of clay entering the cargo the pipe would be raised. Anchor dredging leading to exposure of rock or pure sand strata would theoretically occur more often. Trailer dredging was originally thought to be less of a problem in this respect, as a thin surface layer is skimmed off a wide area, with the final thickness removed being less than the full thickness of workable deposits as determined from original prospecting.

Two facts upset this theory; firstly prospecting only defines the thickness of the workable layer at sampling and profiling points, which are normally too sparse to be adequate (Section 6.2.4); the second is the way in which mining operations are presently carried out. From initial trial dredgings, a series of courses are determined that yield aggregate of required quality, and at any one time only a few courses are used, intensively, to provide aggregate (see photo of side scan sonar record). Using the Decca Navigtator for position fixing, the limited positioning accuracy probably results in the mining of a trench some 50 - 100m wide by, for example, 2 nautical miles (3.6km) in length. Over 240 days (36 hour turnround) a $2500m^3$ capacity vessel would so reduce the level of the seabed by up to 2m. As the working deposit thins, the underlying strata is occasionally seen in the cargo. When the level of contamination approaches a critical level, the course is abandoned and another sought (Section 6.3.1).

Thus, over a period of years, it is possible to surmise that the workable substrate can be seriously thinned and often removed at locations within shallow trough areas. **With the exception of fishermens' reports, the nature of the seabed after a long period of intensive trailer dredging has not been examined. The Consultants are of the opinion that scientific investigation is needed to clarify this matter.**

MAFF support the view that more accurate trailer dredging of deposits in the future should rely on more detailed prospecting data, mining in predetermined patterns using accurate position fixing, and progress monitoring of seabed levels (Section 6.3).

A change in substrate characteristics may also occur as a result of non-removal of oversize material, too large to pass through the draghead trash bars. Consequently, if a deposit which contained substantial quantities of cobbles and boulders were dredged, these would be left in situ, as a residual deposit on the seabed, on the termination of the dredging project.

Finally, a possibly important and unavoidable change in substrate type will occur if a thin layer of material (strictly overburden, but of insignificant thickness) covers the worked deposit. At one end of the scale this may comprise a veneer of shell debris which has taken many hundreds of years to accumulate, possibly playing a local biological role, but not of sufficient importance to prevent mining. At the other extreme live Lithothamnion 'colonies' (a coral-like algae) are closely conserved. In between are examples of little-known importance; Ross (the fishermans' term for Sabellaria spinulosa reefs) is probably commonest. This animal is a tube building worm which lives in colonies, the tubes of which are built of individual sand grains cemented together. They typically inhabit hard ground areas across which some sand is mobile. The reefs may be tens of centimetres thick, and in parts of the southern North Sea, English and Bristol Channels, are very extensive. At present, no evidence exists that dredging has had any significant effect upon the reefs. The colonies play a vital role in stabilising the seabed in high energy, mobile sand environments, the micro-environment so formed containing a highly diverse fauna (George and Warwick, 1986). The Ross reefs are known to play an important role in the food chain, being notable feeding areas for shrimps which in turn are a major food source for commercial fish species. **There is no evidence to support or refute the generally held opinion that the extensive Sabellaria reefs seen today are the result of many tens or hundreds of years of colony building. In the Consultants' view research is required to establish the importance and recovery time of this type of habitat, which, in areas where reefs are extensive but thin, will be at risk from dredging.**

The burial of habitats. The policy of screening out medium and coarse sand at sea, whether on a random redeposition basis as currently occurs or with any attempt to backfill worked out areas (Section 6.3.2), will lead to the burial of originally stable gravel or rock substrates, both within and possibly immediately adjacent to the dredging zone. In most of the dredging areas around the United Kingdom tide and wave energies are sufficient to transport these sands, forming sand megaripples, ribbons and sheets, all largely ecologically sterile substrates (Gray, 1974; Rhoads, 1974). Depending upon the strengths and asymmetries of local tidal streams, much of this sand may over a period of many years be transported away, becoming incorporated in existing sand circulation systems in adjoining areas. The potential for any dredging area to disperse residual medium and coarse sand deposits has never been assessed, nor have any post-dredging studies been undertaken of the extent to which mobile sand deposits blanket original gravel areas.

The fine sand and silt elements of the aggregate deposits, washed overboard with the hopper overflow water, have a greater potential for dispersion. Theoretical fallout and accumulation rates have been calculated for worse-case conditions and are minimal beyond 1km from the dredger, particularly considering the material's resuspension potential. Only one realistic area of general concern has been identified, that of infilling niches in stable substrates immediately adjacent to dredging zones. The potential for this form of habitat destruction can readily be calculated from basic hydrographic data for any specific instances where the proximity of sensitive areas is suspected.

10.2.4 Recolonisation of the seabed

Destruction of the benthos (infauna) results primarily from the direct action of the draghead and pump, with some of the animals remaining in the cargo, but most being returned to the sea as organic detritus. Further mortality occurs as a result of the discharge of rejected sand and fines. Outside the immediate dredging area this effect, in most situations, is thought to be minimal (Sections 10.2.2 & 10.2.3). If a trailer dredger of 2500m³ hopper capacity worked only new ground over a one year period, and mortality was restricted to the area covered by the draghead, 1.3km² of seabed would be sterilised. In reality, trailer dredging is a multi-pass operation, and this figure would normally be reduced by a factor of between 2 & 10. An anchor dredger working deeper deposits would affect an even smaller area.

The point should be stressed that occasional storm-generated events would be expected to affect many benthic populations in water shallower than 30m to a similar degree (De Groot, 1979a). Similarly, other sea uses have comparable detrimental effects, notably dumping, navigation dredging and certain forms of trawling.

Recolonisation takes place from both larval recruitment and the influx of mobile epibenthos from adjacent, unaffected areas. In aggregate dredging, where it would appear that the zone of immediate effect is well contained, the rate of recolonisation should be rapid, with authors reporting partial recolonisation of stable sand and gravel substrates in a matter of months, and full recovery over a period of about two to five years (May, 1973; Kaplan et al, 1975; Augris and Cressard, 1984, quote:)

"In order to establish the optimum size and number of dredge sites for a given

Muddy sand overflow from dredger operating within the Mersey Estuary

region, studies carried out in the Baie de Seine have looked at the time taken in recolonisation. Observation showed this to be rapid. Nevertheless initially the reoccupation of virgin areas is chaotic and depends on larval stock available in the plankton in the months following dredging. Several years are necessary for fluctuations in the fauna to stabilise and produce an equilibrium population."

Comparison of the nature of the benthos between a dredged and a control gravel area (Milner, Dickson and Rolfe, 1977) has shown slight changes in population structure which could not be shown to be related to dredging.

These conclusions relate however to instances where the basic nature of the substrate is not untowardly altered by dredging. Of more critical concern is evaluation of the effect of changed substrate in terms of recolonisation, benthic productivity and the food chain. The replacement of extensive Ross reef with areas of mobile sand is not a temporary but a chronic effect. **It is undeniable that this change of habitat presently occurs. The extent to which it occurs should be assessed, and the implications for local ecosystems established beyond doubt.** Only in the light of this knowledge can the need for increased efficiency in mining practice (as described in Section 6.3) be fully evaluated.

10.3 DREDGING AND FISH STOCKS

10.3.1 Shellfish

Molluscs. (Scallops, whelks, mussels, oysters etc.) Because of the wide distribution of most mollusc stocks, there has been little reported conflict between the aggregate and mollusc dredging industries. Concern would be expressed if a well known settlement and fishing area became liable to an extensive post-dredging change in substrate characteristics.

Crustacea. Crabs, lobsters, shrimps, prawns etc must be caught by aggregate dredgers, but the level of catch is thought to be insignificant in terms of adult populations. The habit of female crabs leaving the main area of commercial stocks to incubate their eggs is of more concern. They congregate in deeper (about 25m) sites with sheltering niches, and pass into a passive state until the crab larvae are released (Howard, 1982). In this state, large numbers of ovigerous crabs in shelters adjacent to a dredging ground may be at risk from a suffocating fallout of fines. At lower levels of fines, though females may survive, disturbance of brooding behaviour

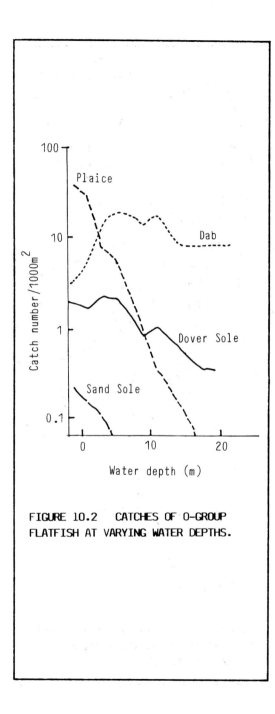

FIGURE 10.2 CATCHES OF 0-GROUP FLATFISH AT VARYING WATER DEPTHS.

may lead to abortion of eggs. This phenomenon is still under investigation at MAFF; if a problem is confirmed care should be taken to identify the existence of proximal 'brooding sites' in relation to modelled fines dispersion and fallout patterns from proposed dredging sites (Section 10.2.1), and to demonstrate the scale of the crab stocks which might be affected in the particular region concerned.

10.3.2   Adult fish

The numbers of adult fish caught by dredging are insignificant in relation to total stocks (J Riley, MAFF, pers.comm.).

The one instance known where adult fish could be affected in very large numbers relates to **sandeels.** These fish bury themselves in the surface sands of the seabed, remaining dormant throughtout much of the winter, only rising into the water column to feed during summertime daylight hours. The sandeel is found extensively throughout UK shelf waters, and is often present in very high numbers in the seabed. Aggregate dredgers which have worked such areas have discovered their cargoes disseminated with rotting fish, and unfit for building purposes. A possible solution (which is currently being tried) is to dredge only during summer daylight hours. Ultimately, if the worked area can be kept sand free, the sandeel population may become displaced into adjacent areas, and conflict of interests should be satisfactorily avoided. As the fish are non-migratory, and fished for use as fertilizer, there is also scope for inter-industry co-operation and the clearance of areas by fishing prior to dredging.

A recent application to dredge in Scottish waters caused fears that the noise of the dredger would upset migratory **salmon.** Acoustic experiments disproved this possibility.

10.3.3   Young fish

Young fish stocks, with small individuals swimming in dense shoals, are likely to be more susceptible to adverse affects resulting from encounters with dredging activity. Fortunately the nursery areas for many 0-group marine fish are found in shallow inshore waters. Figure 10.2 shows the depth/catch number relationships for common commercial flatfish species (from Riley, Symonds and Woolner, 1981). It can be seen that with the exception of Dabs, the young fish are predominantly found inside 15m water depth. The young fish live for 2-3 years in this coastal zone, gradually migrating into deeper waters.

Within the coastal zone itself there are well recognised important nursery areas. Any application to extract aggregate this close to the shore would need to be assessed in relation to these areas.

10.3.4 Spawning

Most fish spawning takes place in the water column, and is neither substrate nor area specific, and hence is unaffected by bottom dredging. Two exceptions are sandeels and herring.

Sandeels. The eggs of this species are laid in the sand during spawning where they adhere to sand grains. Satisfactory development is related to the oxygen concentration of the water in contact with the eggs. Thus smothering of the sand surface and infilling of pore spaces by fine sediment would seriously affect growth of the embryos.

It has already been suggested that dredging may be carried out in a sandeel area without detriment (Section 10.2.3). Blanket objections to gravel dredging from the point of view of effects on sandeel spawning (ICES, 1977) should be considered in relation to the facts that:

(i) Clean sand areas required by sandeels to spawn are necessarily areas of high tidal flow, maximising wide dispersion of the fines element rejected from dredging, and minimising the potential for the persistence of any smothering layer of silt.

(ii) Most spawning would occur on the sand shoal areas in-between the deeps where the gravel basement is exposed. (Figure 10.3). These shoals are aligned parallel to the axis of strong tidal flow, thus plumes generated during extraction would **not** principally affect the bank areas.

(iii) Any minor effect on spawning that did occur would be insignificant in terms of the vast natural stocks of this fish, and on a local scale could soon be made good by lateral displacement from adjacent populations.

To summarise, the sandeel issue, which has possibly been rather overstressed in the past, is one where careful forethought and fisheries/dredging management can in all likelihood eliminate conflict.

Herring. Sites suitable for herring spawning and aggregate extraction coincide in that both dredging companies and the fish seek well-sorted gravel. A stock of herring in balance produces a large surplus of eggs, and therefore limited destruction of spawning grounds as a result of dredging would probably not influence the standing stock. However, in instances of overfishing there is no surplus of eggs, and the issue of spawning ground protection becomes critical. There was a dramatic decline in

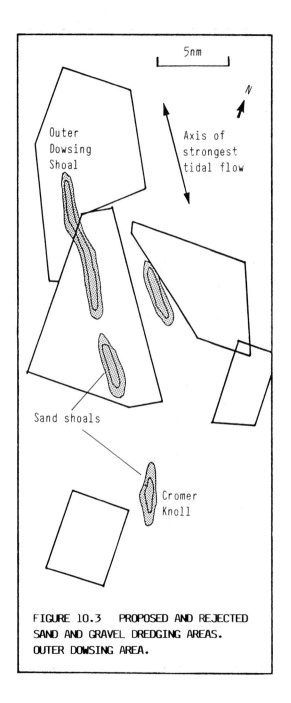

FIGURE 10.3 PROPOSED AND REJECTED SAND AND GRAVEL DREDGING AREAS. OUTER DOWSING AREA.

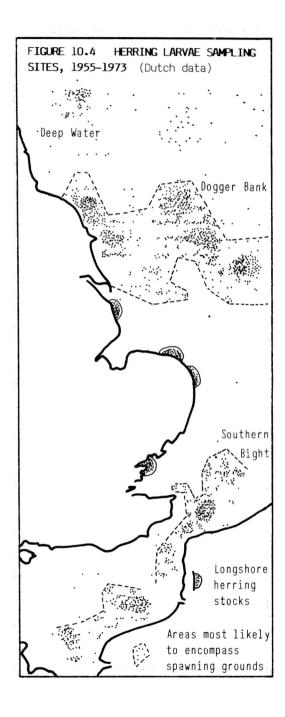

FIGURE 10.4 HERRING LARVAE SAMPLING SITES, 1955-1973 (Dutch data)

European herring stocks during the 1970's as a result of overfishing. The decline in herring stocks was at its lowest ebb in about 1975, but due to the control of the fishery are now actively recovering their former levels with the exception to date of the Dogger Bank stock (A Burd, MAFF, pers. comm.). It is against this background that the potential conflict between aggregate dredging and herring spawning was first discussed (ICES, 1975).

Herring are migratory fish, with the North East Atlantic stock being derived from seven groups based on spawning areas; two of these groups (the Dogger Bank and Southern Bight spawners) lie in the North Sea and English Channel (Figure 10.4). Several small longshore stocks also occur on the East Coast. The herring lays demersal eggs, often in dense mats, which adhere to stones or gravel. The spawning beds are small (eg 3.5km x 0.4km, Bolster and Bridger, 1957) and it is not understood what makes the herring select one locality rather than another. Several facts are clear:

(i) Herring normally spawns on clean gravel.
(ii) Certain specific gravel beds are selected year after year.
(iii) These areas are characterised by low turbidity and moderate to high wave and tidal energy ie free sea water circulation around the eggs is important for oxygen supply and metabolite removal.

Several hypotheses also persist, for which there is little or no evidence.

(iv) The herring returns to the same bank that it originated from in order to spawn.
(v) That the herring can identify its bank of origin through subtle properties of the substrate (eg noise generation under tidal shear, chemosensory perception of chemical and physical attributes of a bank).
(vi) That if a herring cannot return to its bank of origin it will not spawn

The confusion is added to by the difficulty in precisely defining the spawning grounds, due to their small extent and sampling difficulties (Dickson, 1975). As a result plots of larval samples have been relied upon to give approximate locations of the major spawning areas (Figure 10.4). Furthermore, many previous spawning areas in the Dogger Bank will not be utilised until the local stock recovers to its previous level.

As a result of all these considerations MAFF has effectively prohibited gravel dredging from large areas of the southern North Sea.

The Consultants believe that there are responsible ways forward within the constraints that have been defined, eg

(i) No gravel prospecting or extraction should take place in known previous herring spawning areas where stock levels have not yet substantially recovered (areas of the Dogger Bank).

(ii) MAFF should publish the most up-to-date information it possesses on precise spawning areas, sites where freshly released herring larvae have been found, and known previous areas of spawning.

(iii) On the basis of (ii), and avoiding known spawning sites, grounds suspected of containing herring spawning areas could be prospected using coring, grabbing and acoustic techniques only (no bulk sampling).

(iv) When suitable deposits are suspected, hydrographic and coring data should be put together to model spill water plume dispersion patterns. Biologists should then monitor the seabed in the proposed extraction area and the zone critically affected by fines deposition for the presence of herring eggs.

(v) Positive evidence would result in refusal of a licence, or appraisal of the balance of interests in the light of evidence from the following:

(vi) As a matter of priority, the hypotheses that persist in relation to herring spawning behaviour should be examined by a group of experts. Viable ideas should be subjected to experimental tests in the field and laboratory, leading to an assessment of whether gravel areas can be carefully mined and subsequently restored to herring spawning areas.

10.4 POTENTIAL CONFLICT WITH THE INSHORE FISHING INDUSTRY

10.4.1 Description of the inshore fishery

Although the aggregate industry also interfaces with larger, offshore vessels, the increased steaming distances and choice of grounds available together with type of bottom-fishing gear used may account for the lack of reported conflict. It is the inshore fleet of smaller vessels, usually not away from port for more than one day at a time, that has had to accept the realities of aggregate dredging on its doorstep. The ever increasing density of other areas of exclusive seabed usage (Section 8.2, eg pipelines) exacerbates this problem.

In 1981 the total **value of 'coastal' fishery landings** was £36M (Table 10.1; source Pawson and Benford, MAFF, pers. comm.). In that year some 2300 **vessels** ranging from 5-20m in length were engaged full time in inshore fishing (Table 10.2), with 2800 part-time boats. The principle **types of fishing** include otter, beam and mid-water trawling; a variety of set and drift nets; seine netting; hand-lining, long-lining and trolling; potting for whelks, crabs and lobsters and dredging for scallops, oysters, mussels and cockles. With vessel speeds of 5-8 knots, most fishing is done within 12 n miles of the port, allowing daily landings and enabling uncertain weather periods to be worked in safety. In ports where good fishing areas are scarce, these distances may double for the larger boats when a reasonable weather forecast is available. A fisherman may therefore have 40 n miles^2 of attainable grounds, rising to about 250 n miles^2 under maximum steaming distances. Within these radii, there are invariably both good and useless fishing areas.

Organisation. Most operations are run by individuals or families, and there is a considerable amount of seasonal part-time employment. There would appear to be a marked lack of organisation generally in terms of response to external pressures; in worst cases several rival fishermens' associations are present in a single port, or there may be no association at all. In deference to this situation, the MAFF Code of Practice for the extraction of marine aggregates (Appendix 4) nominated the local Sea Fisheries Committees (SFC) to represent the fishing industry. On some coasts at least this has not worked, as the SFC is not regarded as representing fishermen but as a body solely for the enforcement of fishery byelaws within the three nautical mile limit. In Scotland, the Code of Practice refers to local 'fishing organisations' as the point of contact. In the last five years the Yorkshire and Anglia Fish Producers Organisation Ltd has provided a unified voice for the interests of many local Fishermens' Associations between Bridlington and Lowestoft, being represented on the SFCs and maintaining a close contact with the MAFF District Inspectors (M Gowan, pers. comm.). **Most fishermen now belong to a Producer Organisation; it would appear to be in the interests of all inshore fishermen to be represented by an efficient and responsible organisation.**

10.4.2 Operational conflict

A Code of Practice (Section 1.3.1 and Appendix 4) was published in 1981

> "to improve the understanding of the problems and the system of communication between the fishery and marine aggregate dredging industries."

AREA	AGGREGATE REGION	£M VALUE
1. Berwick-Scarborough	N. E. coast	5.40
2. Filey-Hull	Yorks & Humber	1.42
3. Grimsby-Kings Lynn	E. Midlands	1.64
4. Brancaster-Ipswich	E. Anglia	1.87
5. Harwich-Whitstable	S.E. Thames	2.35
6. Thanet-Swanage	English S.E. Channel	2.22
7. Weymouth-Newlyn	English S.W. Channel	15.70
8. Scillies-Bristol	Bristol S.W. Channel	0.30
9. Newport-Barmouth	S. Wales	1.14
10. Portmadoc-Conwy	N. Wales	2.71
11. Rhyl-Duddon	N.W.	0.90
12. Ravenglass-Silloth	N. W. coast	0.51

TABLE 10.1 REGIONAL INSHORE FISHERIES; VALUES OF TOTAL CATCHES, 1981

Source: MAFF

Prospecting. With poor communications and organisation within many parts of the fishing community, it is clearly a problem to pass on details of prospecting programmes to all affected persons, particularly as such information may change on a daily basis. It is also not reasonable to expect fishermen to avoid working in prospecting areas, which are usually necessarily very large. In consequence, operational conflict has and will continue to occur and can only be solved by vigilance and courtesy on the part of both parties. Nearly all fishing vessels are equipped with radio, and dredgers should always be listening on channel 16*. In this way damage to tended gear (eg long-lines) can be avoided. The avoidance of interference with untended gear however (eg sets of pots, trammel nets, Figure 10.5) relies entirely upon the vigilance and knowledge of the dredger's navigation officer. It should be noted, however, that the Code of Practice requires dredging companies to provide Sea Fisheries Committees with a work programme after the licence has been granted and before work commences.

New dredging licences. Under the 'Government View' procedure, local fishermen should be informed immediatly of licence applications, and then have one month to comment to MAFF headquarters via the local DI. Two major problems currently exist regarding the effectiveness of this response.

The first is proving fishing areas. Very few inshore vessels keep authenticated records of their daily landings and precise fishing areas. As a result exaggerated and legitimate claims for interference with livelihood are indistinguishable and objections from fishermen often carry little weight. **It is in the fishing industry's own interests to keep better records in the future; the new EEC Logsheets and Landing Declarations for the collection of fisheries statistics presents an ideal way to record and authenticate such information.** Although this system is only compulsory for vessels >10m, and demands only minimal information regarding species caught and fishing area (ICES rectangle), MAFF are only too willing to receive data from all fishing vessels and in more detailed form. The returns are treated in confidence, with copies being kept by the skipper and the DIs office. Data collation of this type will improve the present situation whereby MAFF often has little alternative than to object to the granting of a dredging licence for vaguely defined fishery reasons. This approach is now seen as producing delay, frustration and ill-will rather than effecting any resolution of the licencing issue.

The second problem is lack of awareness of the hardship caused by displacing fishing areas. Even where good fishing records are kept, it is commonly the case that the income from a few individual fishermen is insignificant compared with the potential returns from aggregate dredging; for example a recent questionnaire relating to a

AREA	VESSEL LENGTH/FULL TIME			PART-TIME VESSELS
	18-12m	12-6m	<6m	
1	111	228	17	440
2	32	41	3	240
3	55	52	6	70
4	17	103	78	356
5	38	110	43	200
6	39	345	85	498
7	118	326	101	60
8	9	98	10	80
9	13	40	6	308
10	11	22	12	23
11	25	76	7	305
12	16	9	8	205

TABLE 10.2 REGIONAL CHARACTERISTICS OF THE INSHORE FISHING FLEET, NUMBERS OF VESSELS.
For areas see Table 10.1

*If radio contact evokes no response, and damage to gear is unavoidable, it is recommended that the caller asks the coastguards (who constantly monitor this channel) to record the call, thus establishing responsibilities for compensation claims. As the Coastguards record all channel 16 broadcasts, this call can be made after the panic of the moment.

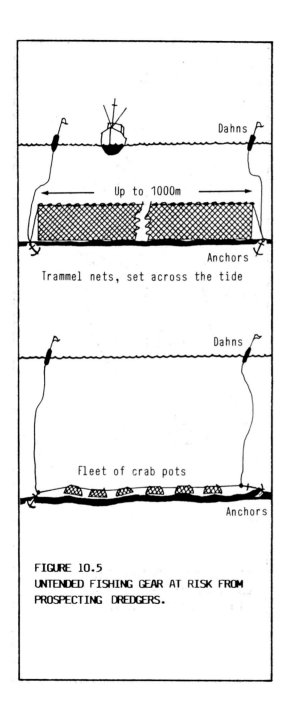

FIGURE 10.5
UNTENDED FISHING GEAR AT RISK FROM PROSPECTING DREDGERS.

proposed East Coast dredging area showed an annual catch value of only £15,000. A dredging licence is not opposed therefore, on the assumption that the fishermen will either take the risk of continuing to work in the dredging grounds, or move elsewhere. As discussed above, the latter option may result in longer steaming times and more down time due to bad weather, affecting the economics of the fishing operation. Furthermore, if worked-out dredging areas prove to be barren (Sections 10.2.3, 10.4.3), the fisherman's catching opportunities will worsen with time. Figure 10.6 shows the juxtaposition of the Lowestoft fleet's available working area and licensed dredging grounds. The remedy to this operational conflict lies in the Industries careful use and refinement of the Code of Practice, to the extent of

(i) **keeping close communication between dredging vessels and the local fishermen, to restrict dredger operations to specific areas on a short term basis, thus allowing unhindered fishing over the majority of the licensed area (the "strip grazing" principle).**

(ii) **mining the ground in such a way (Section 6.3.1) that on cessation of dredging previous fishing areas recover ecologically and become available again to the local fleet.**

In relation to i), the Consultants note that the CEC have recently issued a reminder to the dredging industry of the importance of notifying MAFF of dredging areas which are not currently being worked.

Dredging debris. SAGA Marine and BACMI have recently officially recognised that damage is occasionally caused to fishing gear by debris originating from aggregate dredgers. Screens, pipe joints etc can be accidentally lost overboard during maintenance operations whilst vessels are steaming. A detailed claim form is available; a SAGA/BACMI official will inspect the debris within 14 days and where proven that it is of dredger origin arrange for its disposal, and agreed compensation is to be paid within 2 months.

10.4.3   Long-term sterilisation of bottom fishing grounds

The potential for a worked out dredging area to become recolonised by the benthos and form a productive fish-feeding ground (assuming it to have been so prior to dredging) has already been discussed in Sections 10.2.3 and 10.2.4. It has been seen that some disquiet exists as to whether this end can be widely achieved by current mining practices.

The residual topography of the seabed is a further consideration from the point of view of bottom-fishing (trawling, seining, potting and long-lining). Certain aspects of this problem have been studied in detail at MAFF (Dickson and Lee, 1973). This work has confirmed that most offshore gravel substrates around the United Kingdom are stable; repeated sector scanning and diver surveys in several locations showed anchor dredge pits and trailer dredge furrows to persist over periods of years. These conclusions were supported by measurements of the competence of local tidal currents, which were shown to be incapable of disturbing gravel. Although sand transport was possible, no significant infilling of the dredged areas by sand or mud occurred in the locations studied. Dredged features on gravel substrates are therefore likely to remain for decades. Dredging for sand from more mobile substrates clearly does not pose a problem in this respect.

It is well established that intensive anchor dredging produces a 'moonscape' bathymetry. This prohibits trawling and causes snagging with seining and long-lining, and in consequence MAFF have encouraged trailer dredging (although it should be noted that the potting industry prefers anchor dredging). Trailer dredging is favoured by the dredging industry, as it is operationally more successful in poor weather, and the transition to this type of dredging has gone forward. The way in which an irregular seabed relief may develop under current trailer dredging practices has already been put forward (Section 10.2.3), as has the probability that snagging clay/rock outcrops and residual boulders may also be concentrated in the trough areas. **Such possibilities are scientifically unsubstantiated and should be investigated.** Only when this issue has been clarified can strategies regarding the minimisation of future mining impact be considered.

10.5 THE OUTLOOK FOR THE FUTURE

Three major **areas of concern** have emerged during this study, in the Consultants' views providing the basis for many fishery objections to new licence applications (in relation to the 100 or so Production Licences extant, another 25 have been turned down because of fishery concern):

(i) In contrast to (in most cases) negligible water column effects resulting from aggregate dredging, the residual effect on the substrate, as might be expected from a mining operation, may in some instances be long-term. Such effects however, are likely to be confined to distinct zones within licensed grounds.

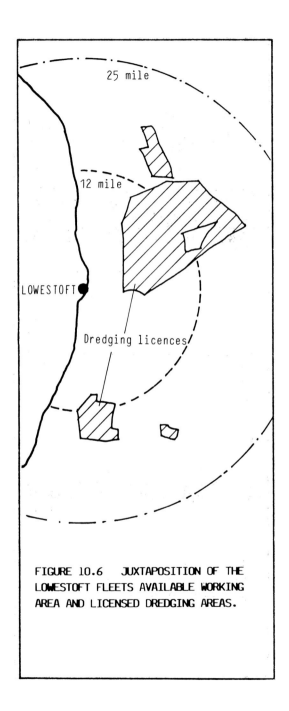

FIGURE 10.6 JUXTAPOSITION OF THE LOWESTOFT FLEETS AVAILABLE WORKING AREA AND LICENSED DREDGING AREAS.

Avoidance of interference with untended fishing gear relies upon the vigilance of the dredger's navigating officer.

(ii) Although dredging is believed by MAFF scientists to have no significant effect on most commercial fish stocks, there are a few exceptions, notably herring and crustacea. The acceptability of dredging in relation to these susceptible species can only be resolved on a **site-specific** basis. Until recently, MAFF has tended to define sensitive areas in rather broad criteria since it lacked either an adequate resource base or good quality prospecting and other information from the Industry on which to refine them; concerns about the ability to monitor accurately the activity of dredgers operating in close proximity to these sensitive sites led also to a tendency to apply large safety zones around them. However, the CEC point out that even for sensitive seabed obstructions such as pipes and cables, a safety zone on either side of no more than 0.5 n miles has been shown to provide an adequate margin of safety for navigational purposes; MAFF note that dredgers are themselves concerned to avoid pipes and cables which may damage gear.

(iii) Dredging grounds tend to be in groups, thus a small number of inshore fishing areas are disproportionately affected. Local discontent has arisen due to displacement for ten years or so to more distant and hence less profitable areas, and also the possibly unjustified worry that worked out areas may be unproductive, and that new dredging grounds may be sought in their vicinity.

There is no single solution to any of these problems, but the Consultants feel that the following lines of development are worthy of consideration;

(i) Increased scientific understanding of a number of issues. Subjects that have been suggested include investigation of residual effects upon habitat within exhausted dredging grounds, the potential role of localised unfishable areas as Fish Reserves, the spawning behaviour of the herring, the behaviour of ovigerous crabs and determination of the long-term economic value of inshore fishing grounds. Appropriate research should be initiated by MAFF.

(ii) Increased data generation during prospecting. Relatively superficial site-specific studies of the natural environment including tides, turbidity, substrates, benthic communities and fishing potential would, when put together with detailed prospecting data, enable better assessment of the acceptablility of dredging proposals (Section 6.2.5). It is important that the scale of environmental investigations that are undertaken are related closely to the expected level of financial return from the proposed dredging project and to problems anticipated.

(iii) Improved accuracy of position-fixing and associated closer control of mining practices as developments in the state-of-the-art permit (including monitoring systems to prevent dredging off station, Section 6.3). This should minimise habitat disruption if fears regarding the effects of trailer dredging prove to be correct, and could lead to the reduction of 'safety margin' areas around zones of known ecological sensitivity.

(iv) Better inter-industry relations along the lines initiated by the Code of Practice, with closer inter-industry contact, allowing fishing of dredging grounds and closed dredging seasons if necessary (ideally liaison officers should be provided by both the dredging and fishing industries).

(v) Improved forward planning of reserve release (Section 6.2.2), to provide more realistic timescales for impact assessments.

It may ultimately be recognised that solutions to some of the problems that give rise to inter-industry conflict can never be realised without affecting the viability of landing marine aggregates. In this event, in terms of the productivity of the environment, MAFF may be required to choose a least damaging option, and there is no reason why both land and sea areas could not be balanced against one another in the same equation. In these circumstances information generated under (i) and (ii) above are necessary as the basis of any assessment, just as soil and land-use surveys are relied upon on land to determine the value of agricultural sites. The issue would be complicated by the need to displace fishing activity; a landowner can sell his land and mineral rights to an aggregate producer, a fisherman cannot.

WHARVES

WHARVES AND WHARF OPERATION

11

CONTENTS:

Wharf Ownership and Control

  Ownership options
  Legislative controls on operations

Locational Criteria

  Site constraints
  Accessibility from the sea
  Accessibility from the land

Shoreside processing

  Cargo receiving and raw material stockpiling

  Processing plant, stockpiling and vehicle loading systems

  On-site value-added plant

Problems associated with wharf operations

  Industrial impacts

  Domestic impacts

  Public open spaces and nature conservation areas

  Harbour navigation

  Summary of wharf impacts

11.1 WHARF OWNERSHIP AND CONTROL

11.1.1 Ownership options

Many sand and gravel wharves are in the freehold ownership of the receiving companies but some are under the control of port, harbour, dock or local authorities. A few are in the ownership of third parties who lease wharves to make best use of their land assets.

Security of tenure is of great concern to operators and even where freehold exists, changes in land-use in an area (eg derelict dockland to residential or recreational pursuits) can lead to local authorities seeking re-location of wharves, if feasible. However, consideration might be given to the strategic position of a wharf both for marketing (road network etc) and of landing materials (dredging cycle).

An operator will prefer to secure a freehold site because on termination of a lease the owner may have other uses for the wharf (hence closure in 1985 of Northwall Quay at Dover and Eling at Totton, Hants). Freehold sites will preferably be outside a registered dock area to avoid unnecessary labour costs (D S Cottell, ARC, <u>pers. comm</u>.) but an operator may opt for a site on 'operational land' to avoid the necessity to apply for planning permission (Section 11.1.2).

11.1.2 Legislative controls on operations

Where wharves for the landing of aggregates are controlled by the local planning authority they will consider need, locational, amenity and road traffic criteria against planning policies. In the absence of planning restrictions, port authorities can exercise 'control' over development through leasing. Port authorities will also be responsible for navigation. Local authority environmental health departments will be able to exercise control on wharves, whether or not planning permission is required. They can consider the impact of noise and dust nuisance, for example. Water pollution of tidal rivers is administred by the Water Authorities.

When planning proposals for new or extended wharves are contentious eg they may be contrary to approved structure plan or other local planning policies, the Secretary of

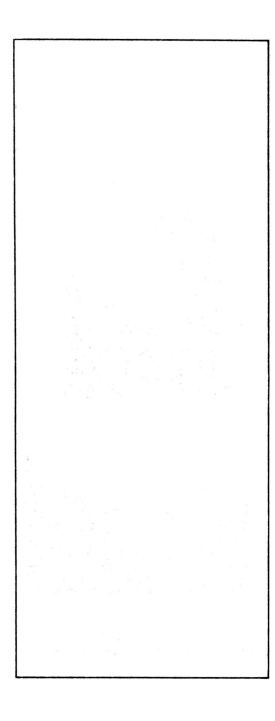

State for the Environment (or Wales or Scotland as appropriate) may call in the case for his decision. Where a local authority refuse an application then the matter can be determined through the appeals procedure. This can involve a public inquiry in which an Inspector appointed by the Secretary of State considers evidence. In the case of minerals and mineral related matters the Inspector often makes his recommendations to the Secretary of State and the latter then determines the appeal. Some appeals are dealt with by the Inspector himself and are not determined by the Secretary of State.

Similarly, appeals against the terms of an enforcement notice served by a planning authority can be dealt with by public inquiry and an Inspector making a report to the Secretary State for the Environment.

In **England and Wales** the erection or extension of wharves for marine sand and gravel off-loading, storage and/or processing require planning permission. They are "county matters" as defined by Section 86 of the Local Government, Planning and Land Act 1980. The Planning Authority responsible for determining applications for aggregate wharves has powers under Sections 87 and 88 of the Town & Country Planning Act 1971 to enforce conditions attached to permissions.

The exceptions to this are wharves (whether leased or not) in the ownership of port, harbour or dock authorities who enjoy permitted development rights under the General Development Order 1977 (Schedule 1, Class XVIII B, 1 and 3). Established user rights might also apply in some cases where a wharf has been used for the landing of any cargoes and a switch made to marine aggregates would not require planning permission. Planning permission would still however be required for the erection of buildings etc not covered by the General Development Order.

There may however, be situations where permitted development might apply to wharf operations under Schedule 1, Class VIII.1 of the Town & Country Planning General Development (Amendment) Order 1981. This relates to, inter alia, any industrial process on land used as a dock, harbour or quay and includes the erection of buildings etc not exceeding 15 metres in height.

In the case of operational land which is sub-let, for example Granville Dock in Dover Harbour - Case Study No 6 - GDO rights do not apply. Planning permission under these circumstances is necessary.

In Greater London, the London Boroughs are the local planning authority responsible for wharf development except in Tower Hamlets where the London Docklands Development Corporation act as the development control authority. Elsewhere in England and Wales the responsible planning authorities are the Shire Counties and the Metropolitan Boroughs.

In **Scotland** similar permitted development rights are conferred by the Scottish GDO (Schedule 1, Class XV B) 1981, but the responsibility for determining whether express planning permission should be granted for development rests with the District Councils except in Highland, Borders, and Dumfries and Galloway Regions where the Regional Councils, and in the Western Isles, Orkney and Shetland where the Island Councils, are general planning authorities.

Where the District Council is the Planning Authority, the appropriate Regional Council may direct that an application should be referred to them instead of being dealt with by the District Planning Authority, if the proposed development does not conform with the Structure Plan or raises a major planning issue of general significance to the District or Regional Planning Authority. The District Council may appeal to the Secretary of State against such a direction, and the Secretary of State (whose decision is final) may determine the appeal. The District or Regional Council may then determine the application.

11.2 LOCATIONAL CRITERIA

11.2.1 Site Constraints

The selection of a new site for a wharf may not necessarily mean an overall increase in marine landings. This is more likely to be part of a company modernisation, re-location or rationalisation of existing operations.

The ideal site is rarely available. In addition to security of tenure (Section 11.1.1), an operator's search for new sites will be infuenced by:

(i) Most importantly, the likely **delivered price** of the products and the relative **location to the market.** Haulage costs are critical since sand and gravel has a high place value. A wharf will more likely be located **within** an area of concentrated demand to offset the advantages of local land-won sand and gravel which will be excavated in a rural area but possibly with a cheaper price ex-pit than the price ex-wharf (Section 4.2.4).

Wharf at Gateshead, located at the centre of the Tyneside conurbation

Dredger negotiating swing bridge in Dover Harbour

Small South Coast wharf

(ii) The **ability of an existing wharf to receive dredgers** without expensive alterations eg piling for jetties. Some ports will be too small physically to accommodate a modern dredger and local planning policies may favour development of the tourist industry rather than trade. Some wharf locations may only be accessible via swing-bridges and/or locked basins which can affect dredger turnround and delivered price.

(iii) In relation to dredger turnround, the **distance from the dredging grounds and tidal cycle** will be important considerations.

(iv) The wharf itself must have **adequate space for stockpiling material** and, where appropriate, **an area for processing plant** to wash, grade and classify dredged material and for the settlement recovery/disposal of silt. Additional space may be advantageous for value-added plant (concrete batching and asphalt plants - Section 11.3.3). An operator will not want to pay for land which is surplus to his requirements.

(v) **Traffic management** is linked with the ability to market (ii above) and may include: back loading (Section 4.2.4) with other goods to make lorry journeys economically viable; lorry departures from wharf timed to avoid rush hour traffic; and vehicle routeing taking into account the general state of the existing road network (avoiding where possible residential areas) and any proposed road schemes or improvements in the forseeable future. In some instances the presence of an existing rail link will be considered for possible long term future expansion.

(vi) In relation to competition from land won sources (i above), the **quality of the dredged aggregate** may be superior to that already available in the market area to be served.

(vii) A local authority's attitude and any **planning policies** that may deter or encourage a new venture.

A suitable site which satisfies all or most of the above points may still be unobtainable, either because the landowner can find a more lucrative use for the wharf (Section 4.2.3) or that the owner has entered into an exclusive-use clause with an existing operator, as happens in some dock areas. The effect of the latter is to preclude other operators from entering into the market place.

11.2.2 Accessibility from the sea

All ports have a **navigation** channel system connecting berths and the open sea. These channels are maintained and buoyed, and movement within the channels is controlled by the Port Authority.

Channels can vary in length from less than one to many kilometres, with corresponding variations in widths, depths and traffic density. For example, Cringle Wharf on the Thames lies 100km upstream from the seaward extent of the Port of London's jurisdiction, whereas in some ports (eg Shoreham) wharves lie within 1km of the bar.

Some channels, like the River Medina on the Isle of Wight or the Langstone Channel in Langstone Harbour, are very shallow even at high water. Passage along such channels may be a constraint. Delays occur due to slow permitted steaming speed, or waiting for traffic movements. While accidents occur, with ships grounding in confined channels or ship collisions, the number is not significantly more than for other shipping movements.

Above a certain size (which varies from port to port), vessels require pilots. For example the threshold requirement is 3500 GRT on London River. Pilotage is decided by individual harbour authorities and is based on: harbour dues, gross registered tonnage or vessel length.

With small wharves where access is along narrow channels, the use of radar for navigation purposes is inappropriate and visual aids are relied on. However when such wharves are fog bound navigation is too dangerous and vessels have either to wait at their berths or anchor at sea until visibility improves eg as at Bristol where fog accumulates especially on Horeshoe Bend of the River Avon and in the River Arun at Littlehampton.

The efficiency of navigation aids and channel dredging varies with the port, and has a bearing on port dues. The simplest ports are usually the cheapest. Dredging of the fairway is sometimes carried out by the aggregate dredging companies.

In addition to channel restrictions, bridges can also present a hazard to navigation. The London River illustrates the problems involved where fixed bridges restrict the passage of dredgers. The only current example above Tower Bridge is at Cringle Wharf, Battersea (Case Study No 1) where specially designed vessels, with the capability of lowering their masts, having a shallow draught and low profile superstructure, enable navigation beneath the bridges.

mv Bowspite, a 'flattie', on the London River

Deep water jetty

Extreme example of spillage, South Coast wharf

Tinsey (1983) highlighted the difficulties over navigating the upper reaches of this commercial tideway, where Westminster Bridge presents the greatest hazard. Vessel timing for arrival at this point, generally confined to a 30-45 minute period, is absolutely critical. This applies especially when going down light, and with the hold flooded with river water provides a clearance of around only 1.7 metres. It is not surprising therefore to find that the majority of the dredger fleet deliveries to the Capital are to wharves located downstream of London Bridge.

Deep water jetties are restricted to the Thames Estuary, Southampton and North East Coast on the Tyne and Tees where demand justifies additional expense. Such jetties support conveyor systems and reception hoppers or pipelines built out to the low-water channel. Alongside facilities are usually limited to mooring dolphins and personnel access with unloading connection. Accessibility is at most states of the tide.

Most other berths in tidal waters are **drying berths.** Most ships are constructed to avoid damage when grounded over the low water period. In a few instances, such as at the Great Western Wharf, Newport there is a problem of foul berths due to rocks etc which preclude vessels lying over low tide.

Drying berths are usually only accessible for a few hours either side of HW, thus restricting turnround time of vessels. This can increase the problem of working on wharves during unsociable hours.

The majority of these berths are situated in relatively confined situations, thus **berthing** problems can be inherent. The turning process often requires warping, anchoring or occasionally a tug. Strong tidal currents can prevent manoeuvring and pin vessels to the berth.

Maintenance dredging of wharf fronts (see also DoE,1986) is often required because of natural siltation, and occasionally by spillage during unloading. Sediment accumulation can form a wedge of material sloping away from a quay. The practise of dredging berths using shore cranes is normally considered unsafe and is usually carried out by the port authority who use their own plant or sub-contract to a specialist dredging contractor. One instance was reported to the Consultants where landings have ceased due to silting of the approach channel to a wharf. Limited financial resources of the harbour authority appears to have been a factor (D P Bown, Dyfed C C, <u>pers. comm</u>.).

Berths with locks are used by dredgers in several 'old' port areas, for example in Liverpool (Wellington Dock), Bristol (Pooles/Hotwells) and Dover (Granville Dock). In these berths ships do not normally take the ground at the wharf unless there has been a build-up of spillage. In some ports lock gates only open for a few hours around high tide, thus restricting access. A further complication arises in Bristol where stop gates are lowered in the Cumberland Basin on spring tides to prevent flooding to the city and significant delays can occur. Locking dues are not necessarily very expensive: the main problem is the delay waiting for traffic movements which often result in a ship missing a tide.

Barge transhipment of sand and gravel along estuaries, rivers and inland waterways is not new. Both marine and land-won material have used this system. Marine aggregates have only been transhipped in relation to specific contracts such as sea defence works along the River Thames. Land-won sand and gravel has become established on a long term basis in certain areas eg Rampton, north Nottinghamshire via inland waterways, and at Mistley in north east Essex where material is barged up the Thames Estuary into London.

It is believed that several companies have considered barging of marine aggregates and have explored the economic and environmental advantages offered by this method of transport. There are nevertheless obstacles to overcome for a successful venture to become established.

One important decision that has to be made is whether the dredged material is taken for **processing at a downstream wharf** whereby the cargo is off-loaded from a dredger, processed and then loaded into sheltered-water barges with onward transhipment to an upstream wharf for stockpiling and sales. The advantages with this system are that it does not delay dredger turnround and obviates the need for processing plant erected in an inner-city area. The disadvantage is of probable duplication of conveyors to run processed aggregate back to the jetties. Additional space would be necessary for stockpiles unless some form of controlled loading could be arranged as each barge moved into position. This method introduces additional costs through double or even treble handling of material and cartage to deliver it to its final landing point.*

The alternative method is to consider **direct transhipment** discharge from dredger to barges at a downstream wharf for **processing at an upstream wharf.** There may in fact be few overall advantages with this system apart from reducing cargo handling costs. The disadvantage appears to be that sufficient barges must be available to accommodate the **total** volume of cargo, every time a dredger berths. The reason is that delays could easily occur to dredger turnround if sufficient barges are not available. Most dredgers would then have to berth on the opposite side to complete its cargo discharge to shore via the luffing boom conveyor. This would be operationally inefficient unless modified so that the luffing boom could discharge on both sides.

Maintenance dredging

* For a period of about 18 months, during 1984/85, cargoes were off-loaded by grab onto a wharf in Southampton which were later re-loaded into a dredger with a smaller draught. The reason for this operation was the lack of an available dredger small enough to negotiate the River Medina on the Isle of Wight. This form of cargo transfer was very expensive and ceased as soon as a suitable vessel became available to land its cargo direct into the Medina.

Lorry loading

In addition, once the barges arrived at their upstream wharf they would then have to be unloaded and the sand and gravel processed. The advantages gained by locating processing plant downstream are lost by this option.

Barge size (capacity) will in part be controlled by local conditions and the ruling craft dimensions for each waterway. Hilling (1981) pointed to a number of factors including origin of materials, their destinations, frequency of traffic, the nature of the commodity and the consignment size. He felt that there were strong incentives to use larger craft particularly as their fuel economics will be reflected in lower tonne-km costs.

The opportunities for barging of marine sand and gravel are more likely to become a reality in London, East Anglia and Humberside rather than for the onward movement of sand in South West and North West England, Wales or Scotland.

11.2.3   Accessibility from the land

Unlike land pits where material can be marketed in effectively any direction, the **market hinterland** of a typical wharf location is restricted because of the coastline. Thus the area of a market is often limited to a semi-circle from a wharf as illustrated by most of the South Coast wharves. This situation does not however extend to many of the major wharves in London, Hull, Newcastle, Liverpool, Bristol and Southampton, where landing points are often many kilometres up an estuary. The use of road bridges and tunnels play an important role in extending the potential areas of penetration for marine dredged material. However, there are also problems associated with tunnels as the London situation demonstrates.

Lorries marketing from wharves at Millwall (Case Studies No 2 A/B) avoid using the Rotherhithe Tunnel because of its tight bends. Wharf traffic will tend to use Tower Bridge and the Blackwall Tunnel. Further downstream, the Dartford Tunnel represents a major bottleneck to north-south traffic using the M25. Wharves located on either side of the river in this area (eg Purfleet and Greenhithe) do not attempt to market on the opposite side of the river because of delays in journey times. The fact that several wharves receive marine supplies on each bank of the river here is also a disincentive to compete across the Thames.

On occasion a congested wharf can be affected by the **uses of adjoining land** eg wharves having a narrow access (which may be shared) and warehouse deliveries at its junction with the public highway can lead to vehicle delays into and out of wharves. Congestion in the adjacent road network and its effects on deliveries must be considered (Section 4.2.4). The ideal situation occurs where immediate access to a major road or motorway can be effected, eg Purfleet (Essex) - A13/M25; Bedhampton (Hants) - A27/A3(M); Fareham (Hants) - A27; Ridham Dock (Kent) - A249; and Howdon (Tyne & Wear) - A1/A187.

With many wharves located close to a strategic road network the use of eight-wheeled vehicles seems likely to continue but many wharves are small, congested and located in urban areas and here the use of four-wheelers may well continue. This will depend to some extent on available vehicles in a company's fleet to serve a particular wharf.

Improvements to the strategic road network in providing by-passes to major towns and cities and the completion of major motorway links, such as the M25, will greatly assist the marketing of marine sand and gravel, not only in reduced journey times to existing outlets but also in providing an opportunity for market penetration into new areas.

Unlike the crushed rock industry where large tonnages are moved annually by rail with an associated high level of capital investment in depots, rolling stock etc, **rail transhipment** of marine sand and gravel is currently limited to two wharves [North Sea Terminal, Cliffe (Case Study No.5) and North Quay, Newhaven (Case Study No.7B)] transporting over relatively short distances (Section 4.2.4).

Possible rail link for an existing wharf operation

The high infrastructure costs, even with grant aid, and a commitment by an operator when entering into a Freight Agreement with British Rail for the movement of a guaranteed minimum annual tonnage over a ten year period, does not lend itself to a major switch of road to rail traffic. Only in a few situations where a future market is more or less assured, as in London, will any new rail links be established. The existing rail linked wharves may increase rail movements depending on competition with other supplies. This is important as high utilisation of wagons is necessary to minimise their amortised costs.

Following a re-structuring of their Freight Department in September 1984, British Rail are looking to be more efficient and to offer existing and potential customers a quicker response to possible freight movements and costings. It is to the Industry's advantage to provide British Rail, at an early stage of wharf development or expansion, details of projected markets, anticipated tonnages, availability of customers' wagons where appropriate, and any large fluctuations in sales due to major contracts. The most helpful response will come if such details are furnished. Possible freight rates can then be assessed: lower rates can be negotiated for large tonnages and long term commitment.

Self-discharge scraper buckets

Shoreside receiving hopper

It is however important to appreciate that a rail linked wharf is not in itself any guarantee that marine landings will be transported by rail. There have to be established receiving depots in, or immediately adjacent to, areas of demand and this should form part of an operator's marketing package.

11.3 SHORESIDE PROCESSING

11.3.1 Cargo receiving and raw material stockpiling

The present systems for discharging a dredger's cargo involves either shipboard or shoreside plant. The use of drag-scrapers, grabs, rotary bucket wheels and conveyors or pumps on a vessel (DoE, 1986) or grabs, cranes or shovels on a wharf seem little likely to change over the next sixteen years (ie up to the period ending 2001). Although 70% of wharves in the UK use grab discharge as the means for cargo landing, the tonnage by this method represents only some 37% of the total marine landings.

Typically, discharge rates are 200t/hr for grab and 1500t/hr for self discharge. In the former, wide variations occur depending on the size of clam shell eg 120t/hr - 285t/hr.

In isolated situations (as at Hotwells, Bristol) the introduction of a wharf mounted gantry crane, which runs out over the dredgers's hold, does not require dock labour. A move in that direction by a few small wharves to reduce labour costs might occur only where existing cranes need replacing. The economics of such cranes has to be weighed against speed of discharge since the grab can have a smaller capacity than that of a conventional wharf mounted grab crane.

The move towards self discharge methods will greatly assist turnround times where deep water berths exist ie where vessels can sail out at any state of the tide. Production economics will however, only marginally change.

Most self-discharge conveyor systems feed into receiving hoppers which are fixed on the wharf. The ability to supply material to these hoppers is dependent on the flexibility of the dredger's luffing boom. In a few instances (eg Purfleet and Greenhithe) reception hoppers have a variable level control to adjust to the ship's transverse conveyor with the particular state of the tide.

Landed material is placed in raw stocking areas either by grab-crane off loading into storage bins or pens or by grab/self discharge means onto an overhead conveyor for distributing material to the desired storage bays/pens below.

With pump discharge systems (DoE, 1986) river water is pumped aboard a dredger and sprayed as water jets from valves near the bottom of the hold. This slurries the cargo which moves towards the centre of the hold where it flows into a duct below. The cargo is then pumped ashore along a rigid pipeline (unless site reclamation works by direct placement are being undertaken where flexible pipeline to the shore is used).

Once ashore the cargo is discharged into a reception tank or tanks where excess water is weired off and allowed to flow back to the river after settlement of suspended solids. The tanks, which can hold around 2-5000 tonnes, are dug with an excavator to supply raw feed to a processing plant. Where processing is not involved the excavated material is stocked ready for sale.

On occasion the rate of discharge from the vessel can exceed the rate at which the reception tanks can be cleared of aggregate. Under these conditions the dredger has to temporarily cease discharging its cargo until sufficient spare capacity has been created by stocking some of the raw material to one side of the tanks. Sand will be pumped at a higher rate than sand and gravel and the stonier a cargo the longer it will take to discharge. Typical discharge rates are 4000t/hr for sand and 1250t/hr for sand and gravel using the same vessel.

'As dredged' material, which is not processed on-shore, will normally have been screened at sea to approximate an overall grading requirement.

In the Bristol Channel and North West England the landed cargo is only sand. Screened cargoes are off-loaded into stocking pens, depending on grading, to await sale. Some of the finer grained materials will be used as building sand (for example from the Denny Shoal, off Portishead) or coarser sands used in concrete manufacture, such as from the MacKenzie Bank (by Flat Holm Island), both in the Bristol Channel.

'As dredged' cargoes are used for fill/reclamation projects, normally by pumping directly into a site, as for example in the Port Glasgow dock improvement and at Stone Marshes, Kent.

11.3.2 Processing plant, stockpiling and vehicle loading systems

The washing, grading and crushing of aggregates, even after ship-board screening, takes place along the East Coast, Thames Estuary and South Coast for the majority of the sand and gravel landed. Screening at sea only approximates the customer's product size requirements of the coarse: fine aggregate ratio (Section 7.1.2).

Pump discharge pierhead connections

Excavating holding tank during pump discharge

Crushing plant

Processing plant with stocking out conveyors

Inevitably reject material (+40mm) will occur in some cargoes, particularly from certain banks, and the installation of crushers is a general necessity. Similarly, the installation of metal detectors at most plant sites is a safety requirement in case of ammunition fragments etc.

Shell is not a major problem (Section 7.2.1).

Processing plant will be designed for each wharf, taking into account the characteristics of the material to be landed (eg a high sand content will necessitate additional treatment plant).

Apart from an assessment of the nature of the raw material, information on the desired end products and anticipated output are necessary design criteria, linked with the availability of stocking areas, water supply and disposal of wastes.

In a gravel pit a processing plant has a raw feed which is generally continuous, being linked with a working face a short distance away. For wharves it is vitally important to have adequate stockpiling areas for 'as dredged' cargoes otherwise if the raw supplies are not maintained due to bad weather, then the wharf will have to cut back on production. High production rates ie the hourly rated **capacity of the processing plant** are therefore necessary to meet a sudden demand where there is limited space for processed products. The tendency to seek high capacity processing plant is to enable short term peaks in demand to be met, unlike plants located in gravel pits which essentially rely on a more or less continuous raw feed of material to build up adequate processed stocks. The rated capacity of wharf plant must therefore be regarded with caution as the theoretical maximum annual throughput is unlikely to be achieved.

The hourly rate of processing plants used on wharves varies from 70-400 tonnes, the smaller capacities being located generally along the South Coast.

The size of raw stockpiles will vary from wharf to wharf depending on available space but on the South Coast, for example, these will often be small since vessels have short steaming distances from dredging grounds and operate in relatively sheltered waters. Stockpiles of processed material are usually larger. The converse situation applies generally along the River Thames and North East Coast where larger raw stocks are held on wharves because of adverse dredging conditions in the North Sea and because greater steaming distances are involved.

Re-handling of raw stocks for processing may either be achieved by loading shovel, or crane grab feeding a hopper conveying material to the plant, or by underground conveyor beneath a surge pile. Examples of the former are at River Road, Littlehampton and Cargo Fleet, Millwall and of the latter type, at Baker's Wharf, Southampton and Cringle, Battersea.

Processing (Section 7.1.1) is carried out using washing and grading techniques involving high pressure water sprays directed onto multi-deck vibrating screens of varying mesh size to produce clean coarse aggregate @ 40mm, 20mm, 10mm and 5mm. Where oversize material is likely (+40mm) a crusher may be installed so that smaller material can be returned to the screens for grading. Oversize material is sometimes retained in stock for special contracts, including minor beach nourishment schemes. Unlike many land operations, it is not necessary to install a washing barrel to process marine aggregate. With large throughput tonnages it is sometimes necessary to duplicate vibrating screens, as at Burnley Wharf, Southampton and North Sea Terminal, Cliffe. The latter also has sand towers and dewatering screens in triplicate.

Washwater is usually from a mains supply and the system is topped up/drained as necessary having regard to monitored chloride levels. Occasionally lake or even river water is used where the chloride concentrations are acceptably low.

Fine aggregate (<5mm) is passed to sand classification plant for dewatering and removal of silt. Dewatering of sand may involve one of four methods. That which is used most commonly on wharves is the tower mounted hydro-cyclone with optional dewatering screen. Other methods are sometimes used. Sand classification occurs where two grades are required: one for use in concreting and the other in building or asphalting. Similar plant to dewatering is used in classification: the centifugal system is again most commonly used.

With space often at a premium on a wharf, the amount of silt for disposal has to be kept to a minimum, and this also avoids the need for frequent off-site emptying to a local landfill site. Some fines are blended back into the sand within the limits set down by the British Standards (Section 7.2.1). To reduce off-site disposal the use of more precise classification to recover additional fine sand and silt is sometimes adopted, as at Union Wharf, Millwall.

Processed material can be held in overhead bins with either direct discharge into waiting lorries below or by conveyor transfer when limited turning space is available for lorries. If space permits, the plant may have stocking-out conveyors to ground storage where material is either taken by loading shovel to a lorry or by underground conveyor recovery as stocking-in conveyors to overhead bins. Overhead bins with

Underground conveyor from surge pile

Centrifugal separator and dewatering

Processed material in overhead bins

Concrete batching plant

Asphalt plant

conveyor feed to lorries can be found at North Quay, Newhaven (Case Study No 7B) and Howdon Wharf, North Shields. Overhead bins with direct lorry discharge operate at Northfleet and Cringle Wharves. Purfleet and Kendalls Wharves both have ground storage with loading shovels; and Cliffe has stocking-in conveyors.

At Cliffe, where rail links operate, loading of wagons is done by an overhead travelling tripper which can load either side at a rate of 6½ tonnes per minute on two 128 metre sidings. At Newhaven a loading shovel takes graded material from stocking pens for discharge into railway wagons along a 360 metre siding.

11.3.3   On-site value-added plant

Many of the wharves receiving sand and gravel have concrete batching plants located on-site and the bulk of processed material leaving the wharf is also supplied to similar plants in the market area. In a few instances, eg at Purfleet, marine sand and gravel is also used in asphalt plants. The installation of value-added plant provides a greater scope for ex-wharf sales (Section 7.1.1).

11.4   PROBLEMS ASSOCIATED WITH WHARF OPERATIONS

11.4.1   Industrial Impacts

Generally the location of an aggregate wharf within an industrial setting should not give rise to problems. Occasionally there may be conflict or potential conflict where an adjacent operation could be affected, as at Birkenhead Dock, Wirral which adjoins a flour mill where blown sand from stockpiles has been controlled by a water sprinkler system to keep the sand at its optimum moisture content. Different forms of control measures may be necessary at other wharves.

The converse situation can also apply whereby another user (Section 11.2.3) might affect the quality of aggregate by wind blown dust. An example of this is illustrated by one producer sharing a wharf with a coal operation in North East England. The aggregate producer decided to withdraw from the site and to relocate elsewhere for fear of coal dust contamination in concrete. Often however wharves are in the freehold ownership of the operating companies who are therefore loath to give up sites, accepting that some inconvenience to themselves has to be tolerated.

11.4.2 Domestic Impacts

Some wharves lie very close to housing areas, either because of historic association or through re-development of old dock areas, with the result that noise, dust, fumes and traffic can cause a nuisance, especially when working occurs at unsociable hours. Some wharf operations are not restricted on their hours of working and given sufficient demand can work either a two or three shift system. In the current economic climate only a few wharves operate throughout the night, especially as this could give rise to complaints from local residents. An example of a wharf capable of three shifts is Northfleet on Botany Marshes, Kent (Case Study No 4).

Objections are often likely to any intensification of wharf activity through the installation of value-added plant at these wharves. Tolerance is very dependent on the local prosperity of the area and the demand for jobs (some planning authorities have policies encouraging development which produces local employment).

The proximity of a wharf to other forms of development may lead to conflicts with planning and environmental health authorities. Informal agreements or voluntary codes of practice are an effective means of control.

Generally, domestic impacts can sometimes be minimised by limiting the hours of working and production, thus restricting the potential number and time of lorry movements through residential areas. Note however that in some small ports, if supplies are to be maintained, problems of this kind must be endured. Although the situation of these wharves is not ideal from any point of view, some will in the course of time close due to relocation and rationalisation of operations. An example of this is Bath Lane Quay at Fareham, Hants which was located in a residential area: as a small wharf lorries had to queue in the adjoining streets. Vessel unloading was forbidden at night by notices served under the Control of Pollution Act 1974. Noise and dust was also a problem. That site was relocated in 1984 to Upper Quay (Case Study No 10) and the benefit of **existing use rights** whereby only limited control can be exercised. New wharves are unlikely to be permitted in residential areas.

To ameliorate the problem of noise emissions many processing plants are now fitted with polyurethane decks and all chutes having rubberised linings. A plant itself may be clad in sheet metal. At Kingston Railway Wharf, Shoreham, 180 metres from houses on the opposite side of the River Adur, the plant has been fitted with a rubber noise-suppression cover around the vibrating screen; a 5 metre high reinforced concrete wall built around the crusher unit; and a 4 metre high wall contains the concrete batching plant loading area (Anon, 1983).

Residential development close to a wharf

A corner of Langstone Harbour

Traffic control measures may be necessary where the road network of an area is sub-standard or there are problems with busy approach roads. These may be either self-imposed by agreement on routeing and/or vehicle numbers, or formally by a local authority Agreement under Section 52 of the Town and County Planning Act 1971 [or Section 50 of the Town and County Planning (Scotland) Act 1972]. Although such agreements will allay some local fears, doubts have been expressed to the Consultants as to their enforceability, particularly where contractors' vehicles are involved. Avoidance of rush hour traffic and back loading are measures which an operator will take into consideration when planning an efficient use of his road transport.

Overall, traffic management measures are unlikely to affect the existing throughput of wharves. Improvements here may come about as local authorities implement new traffic schemes in urban situations. Where access to some wharves is difficult a local authority may have a mineral policy precluding further expansion, eg at Bath Lane Quay, Fareham.

11.4.3 Public open spaces and nature conservation areas

Pollution, for example noise from a processing plant or cement dust from an overblow in delivery at a concrete batching plant, is considered a threat to nature conservation areas though no supporting research has been found. Noise can travel over large distances, depending on climatic conditions, local terrain, development and the prevailing ambient noise levels.

Visual amenity can be protected by planting around the perimeter of a wharf set in a rural location. Alternatively, concrete walls as indicated in Section 11.4.2 may prove acceptable.

To attempt to totally blot out a wharf from public view is not necessary: in the Consultants' experience **most people like to see shipping activity** (berthing, unloading and sailing) **as an indication of an area's prosperity and as an interesting, traditional spectacle.**

There is little interference with water sports etc beyond that which is necessary for normal dredger safety considerations. The main objections seem to come from popular yachting areas, such as at Ramsgate marina where the entry of dredgers and establishment of new wharves has been strongly discouraged by the port authority. In other areas where yachts are in abundance dredgers do not create a conflict in waterspace usage, as in Poole Harbour and Langstone Harbour. In fact **in the latter area the presence of Kendalls Wharf is an advantage to a local sailing school** as it prevents the proliferation of moorings which would otherwise hinder sailing instruction, **and** by the same token reduces **pressure on over-exploitation by the public of an SSSI.**

11.4.4   Harbour Navigation

In commercial ports dredger activity is normal and there is usually no conflict with other users. In yachting harbours the close proximity with a marina should be avoided wherever possible to reduce risk of collision. On the fringes of commercial ports where recreational development has taken off in recent years, such as along the River Itchen in Southampton, problems have occurred. At one marina a landing stage was built out half way across a dredger's berth without approval although subsequently removed.

11.4.5   Summary of Wharf Impacts

In general there would appear to be little or no adverse affect of landing marine sand and gravel. This is because the vast majority of wharves are located either in traditional dockland areas or are remote from human habitation. Occasionally conflicts do arise sometimes to the detriment of the aggregate producer with other industrial operations nearby (eg flour mills or coal yards) but more often to the detriment of local inhabitants or amenities through noise of plant or machinery and lorry movements through residential areas.

Marina adjoining a South Coast wharf

PROSPECTS FOR THE INDUSTRY

REGIONAL ASSESSMENTS

12

CONTENTS:

The model for prediction

South East

East Anglia

East Midlands

West Midlands

South West

North West

Yorkshire & Humberside

Northern

South Wales

North Wales

Scotland

12.1 THE MODEL FOR PREDICTION

The Consultants have used the following model for prediction of the contribution that may be made by marine sand and gravel to the UK market.

12.1.1 The demand for marine aggregate

The work of the Regional Aggregate Working Parties (RAWP) formed the basis for the preparation of the Government's 'National Guidelines' (DoE Circular 21/82, Welsh Office Circular 30/82). For the purposes of this report, the model for prediction of likely regional trends is based on that Circular together with the 'National Planning Guidelines' for Scotland.

The current forecasts of aggregate demand are those prepared by the DoE in July 1981 (Table B1 of 'National Guidelines'). These cover the period 1981-1991. However, under the terms of reference for this study consideration has to be given to 1996 (to coincide with the review date for county structure plans). Since the DoE has not produced any new demand forecasts, the Consultants have assumed that the levels indicated in the 'National Guidelines' will not materially change between 1991 and 1996.

Each of the RAWP's in England and Wales considered a Basic Option for their region, which assumed the same proportions of aggregate (in their various forms) that were provided in 1977 would continue to meet total demand between 1981 and 1991, with more general observations on the situation to 2001. Where the implications of the Basic Option were unacceptable to RAWP's, slight variants were adopted. The only significant departure from this was in the South East where the Basic Option was rejected and in its place a general philosophy was submitted.

The strategy finally adopted for each region was the **Preferred Option.**

The background to demand is set for each region by broadly describing the relation of population centres to existing or potential wharves (details of the latter being largely derived from local planning authority and port authority questionnaire returns). The interaction of land-won sand and gravel, crushed rock and marine supplies is summarised where relevant for each region.

FIGURE 12.1
AREAS OF REGIONAL
AGGREGATE WORKING PARTIES

It should be noted that all planning predictions and forecasts, although necessary for guidance purposes, may not be realised in practice and can be amended by Government policies or the general economic situation.

12.1.2   Analysis of the effects of constraints on the supply of marine material required in the preferred option

An initial premise is made that the percentage of demand supplied from the sea will not change within each region from that specified either within the Basic Option or the level of demand prevailing in 1983 if greater ie no growth of the marine sector. The supply or marine material is then examined in the light of known future constraints on licensed reserves and plant. In considering each region regard has been taken of

(i)   the information supplied by all maritime mineral planning authorities in England and Wales and the maritime planning authorities in Scotland on wharves and planning attitudes, or policies where applicable.
(ii) The capacities of the dredger fleet, and
(iii) the availability of licensed offshore reserves adjoining the region.

From this stage the necessity for increased throughput from plant is looked at, firstly in terms of possible efficiency increase, and then in terms of investment in ships, etc. and the availability of new wharf sites. Shortages in licensed reserves are looked at from the viewpoints of transport from more distant grounds, opening of prospected reserves and the potential for undiscovered resources.

12.1.3   Examination of the possibilities of changed landing patterns

As a second step the possibility that the percentage of demand supplied from the sea will increase regionally, as a function of exhaustion of land deposits, changing market conditions or policy, is investigated.

Where projected shortfalls in demand for land-won sand and gravel have been identified in the 'National Guidelines', comments are made on the possible supply of marine aggregates into that region. The implications of this are that, provided the necessary wharfage and infra-structure are available, then the level of commitment by local planning authorities over the release of land for sand and gravel working may be reduced. (Note: in an expanding market both the continued level of land released **and** increases in marine landings could occur). Revised predictions of marine supply are then related to existing licensed reserves and plant capacity where feasible.

12.1.4   General points

In considering each part of 'mainland' UK, divisions have been based on the AWP areas and Scotland, as shown in Figure 12.1.

Vessel under-utilisation has been calculated as a percentage in some market areas. This has only been possible where a small number of vessels of known cycle time normally operate into one or two regions. Efficiency is expressed relative to 80% utilisation of available time, which is the best that can normally be achieved (Section 4.2.2).

It has not been possible with the information available to make allowance for proposed **major construction projects** such as the Severn, Morecambe Bay or Wash Barrages or Sizewell B Power Station. The implications of the English Channel Fixed Link are considered in Section 12.2.4. Timing, duration of operations and quantities are uncertain. Should any of these projects commence during the period under review (to 2001) then undoubtedly they will have significant implications for the marine sand and gravel operators who are likely to be well placed to supply the large quantities of construction materials required. In addition, the construction of the Severn Barrage could have adverse implications on some of the existing dredging grounds in the Bristol Channel through the obliteration of some licensed areas and cutting off the natural replenishment of others (F G Parrish, CEC, pers. comm.).

Proposals are underway to bring sea-borne granite from the west coast of Scotland, planned to commence in the late 1980's, to Liverpool, the Royal Portbury Dock in Bristol, Southampton Water, and Purfleet on the Thames Estuary. The granite is a high quality roadstone and its main outlet initially is expected to be existing (and proposed) coating plants. To what extent such material will be used in competition with marine sand and gravel is unknown at this stage.

12.2   SOUTH EAST

12.2.1   The potential for marine aggregate landings

The region comprises a northern sector containing the counties of Oxfordshire, Buckinghamshire, Bedfordshire, Hertfordshire, Essex and Greater London (with marine supplies coming largely from the Thames Estuary and East Coast), and a southern sector containing Hampshire, Berkshire, Surrey, West Sussex, East Sussex and the Isle of Wight

---

Table Symbols

A = high level of encouragement from port
B = marine landings would be encouraged.
D = no policy
(L) = locked
(T) = tidal
L = length ⎫
B = beam   ⎬ max. vessel dimensions
D = draft   ⎭

Figure Symbols

HULL   Existing ports landing marine aggregates
3.   Potential landing ports (see accompanying table)

Population Density

⬚ <40 per km^2
▨ 40-200 per km^2
▦ 200-500 per km^2
■ >500 per km^2

Source of population data: Complete Atlas of the British Isles, Reader's Digest Association, 1965.

TABLE 12.1   EXPLANATION OF SYMBOLS USED IN THE TABLES AND FIGURES SHOWN IN CHAPTER 12

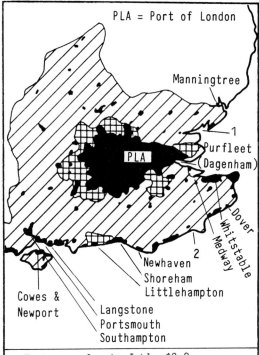

FIGURE 12.2   THE SOUTH EAST

1. COLCHESTER AND BRIGHTLINGSEA
2. RYE

TABLE 12.2   POTENTIAL LANDING SITES IN THE SOUTH EAST

(with supplies coming largely from the South Coast). Kent occupies an intermediary position with supplies from the South Coast, Thames Estuary and East Coast areas.

The Regional Aggregates Group (RAG) rejected the Basic Option in view of what they saw as major changes in the regional supply pattern since 1977 and how this was likely to continue. They were influenced by:

(i) investment plans by both British Rail and the Industry for imported material
(ii) the marine dredging sector of the Industry anticipating further expansion
(iii) some areas within the region being unable to maintain their 1977 proportional contribution to supply without relaxation of policies on high grade agricultural land and attitudes towards the environment.

A policy of gradual run-down in land-workings was felt by RAG to be preferable to the sudden adjustment which would otherwise be necessary in some areas by the end of the 1990's when potential reserves would be exhausted.

RAG acknowledged that there were uncertainties relating to the expansion of some sectors, such as marine dredging. The **Preferred Option** was couched as a general 'philosophy' which encouraged the aggregates industry's intention to increase its contribution from marine, rail-borne and sea-borne sources. Planning permissions would be granted by mineral planning authorities to meet demands that could not reasonably be met from these sources. A continuing reduction in land-won materials as a proportion of total demand was expected.

The **National Guidelines** indicate that with the persistence of defined trends, 20% (6.5-8.7M tonnes per year) of total production to 1991 should be sought from marine dredged sand and gravel. Planning permissions should be ensured to permit 75% of production to come from land-won sources, 2% from crushed rock and 3% from other materials produced within the region. Imports of 8-10.5M tonnes per year (mainly of crushed rock) also would be necessary to meet aggregate demand within the region, coming principally from the South West and East Midlands.

There were forty-two wharves in the region receiving marine sand and gravel in June 1985 (Figure 12.2). In addition, one company is proposing a new wharf for the River Thames at Dagenham. The quantity of marine aggregates landed in the region far exceeds the landings in any other region of the UK; production has increased steadily to around 8M tonnes. The most important areas for landings are the eastern half of Greater London, Kent and Hampshire.

In terms of **market accessibility** the region has a population in the order of 12M persons, probably more than 50% of which lies within 50km of wharves currently receiving marine aggregates.

Aggregate quality requirements from marine landings are for concreting sands and gravels and building sands.

12.2.2   Constraints on marine supplies, northern sector, 1986-1996

The London, Essex and Kent areas are principally supplied from the East Coast and Thames dredging grounds (Figure 4.1). **Licensed reserves** in these two areas are 41 and 15M tonnes respectively; four or five licence applications are being considered and further potentially workable deposits are known to exist within the area, although their use is possibly constrained by coastal stability or fishery problems. Reserves in these grounds are currently depleted by about 7.5M tonnes per year, which in view of continuing trends in the South East region and projected requirements of the East Anglian region (Section 12.3) may be expected to increase to a least 8M tonnes per year before 1996. This production contains a substantial (2.9M tonnes per year) element used for export. It is therefore apparent that reserves supplying the South East could be exhausted by 1993. In reality considerable problems are already being faced by some companies, particularly in the Thames Estuary where most licences are essentially exhausted.

Vessel availability is not generally a constraint, as several dredgers operating on the East Coast are currently laid-up or significantly underutilised, as emphasised by the level of exports and the active quest for new markets along the North and East Coasts. Certain vessels, notably the 'flatties' that supply the Central London wharf, will require to be replaced in the early Nineties, but the remainder of the thirteen vessels regularly employed in this trade will be active until 1996-2005.

Although there is little scope for new **landing sites** within the area, with the possible exception of the Colchester/Brightlingsea area, there is general encouragement of further facilities in ports already accepting aggregates (eg Tilbury, Canvey Island (west), Sheerness, Ridham, Beckton and London River wharves). It is thought that existing shore-plant could cope with a further 2.0M tonnes per year throughput.

It is clear that the major constraint in fulfilling the Preferred Option in this sector of the South East region is lack of licensed reserves. Exports to the Continent could seriously affect the prospects for increased landings in the South East region although at present they are relied upon by the Industry to maintain operations underutilised

> *During the five years 1981-85 inclusive the quantity landed in the UK has increased, with a marked rise in 1985 to 14.0Mt. Exports to the Continent have fallen by around ½Mt to 2.5Mt in 1985.

nationally because of the recession; projected expansion of UK marine landings will hopefully replace exports.* Substantial new reserves MUST be released within the next few years in areas accessible from the Thames Estuary ports. Resolution of fisheries and coastal stability issues is looked for, which could release around 50M tonnes of reserves (of which probably just over half would be derived from South Coast grounds). Prospecting initiative to identify and justify the release of further reserves is vital for the medium to long-term if marine landings are to increase in accord with the regional guidelines (Section 12.2.4).

12.2.3 Constraints on marine supplies, southern sector, 1986-1996

South Coast reserves are the primary source of marine aggregates marketed in this region. Some 41M tonnes are licensed for dredging, and 60M tonnes are pending government decision (note the bulk of the latter deposits are in the Eastern Channel, and would be shared with, and possibly even primarily used by, the northern sector market; this has been taken into account in the previous section). It is generally felt that the South Coast between Exeter and Dungeness may offer scope for further discovery of reserves, particularly in water just deeper than that currently dredged, and accessible to the next generation of vessels.

In 1983 the rate of depletion of South Coast reserves was 3.8M tonnes per year; this might conceivably rise to 5 or 5.5M tonnes per year by 1996. At a simplistic level, considering existing reserves and reserves liable to be released in the coming five years, reserve exhaustion should not occur prior to 1996, although individual company shortages may cause supply problems, which can only be solved by prospecting for new reserves.

All suitable ports currently receive dredged aggregates, with the possible exception of Rye. There is some scope for relocation and new wharf development within the major ports. It is thought that existing facilities could cope with an additional throughput of some 1.2-1.5M tonnes per year.

The Consultants consider that a major programme of rationalisation will be required in relation to the **fleet serving the South Coast,** normally comprising some 17 ships. Before 1990 five to ten vessels of 500-1000 tonne capacity, mostly, but not all sheltered water craft, will probably need to be retired, and by 1996 a further five 800-1500 tonne vessels will have reached the end of their expected working life. If it remains economically unattractive to build vessels of less than 4,500 cargo tonnes (Section 4.2.2), a major rethink of the landing strategy along this coast will be

the next year or two, effectively doubling this landing figure. The marine contribution to the region's production during the decade 1986-1996 is therefore more realistically assumed to be 0.6M tonnes per year.

In the Consultants' view the initial success of the new landing ventures in the area will rely heavily upon the immediate proximity of the dredging grounds, allowing a 12 hour cycle time and hence keenly competitive prices. Thus it is assumed that the bulk of the **reserves** supplying this market will be derived from the East Coast grounds, where some 41M tonnes of material suitable for concreting aggregates are licensed. These reserves however are presently the principle source of aggregate for the London and export markets, and are being depleted at a rate of 4-5M tonnes per year (2M tonnes per year for export). At this extraction rate the local availability of material for the East Anglian region will start to be problematic soon after 1990, and it appears that licensed reserves will be exhausted by the mid 1990's. Four reserves are being considered for licence application, and some 18M tonnes have been refused in the past due principally to fisheries objections. Future prospecting in the immediate area may identify further workable deposits.

Vessels supplying the region will not need to be replaced prior to 1996.

Wharf capacity is not thought to be a constraint, as exemplified by current and proposed operations on the East Coast of the region. Wharfing facilities are also available in the north Norfolk/Wash area (Table 12.3).

The continuity of offshore reserves post 1990-1994 is the major constraint on future landings in the East Anglian region. It is thought that proximity of dredging grounds is of major economic importance during the initial period of marine penetration to this region, improving competitiveness with land-won material. One means of ensuring adequate reserves is to divert some of the London and export demand to other grounds before the East Coast reserves become too depleted.

12.3.3 Changed landing patterns 1986-1996

The release of reserves of land-won sand and gravel is foreseen as a problem post 1991 given current planning policies, and there may therefore be some scope for increased marine landings within the region. Such increase will be controlled by competition with crushed rock supplies entering the region via its landward boundaries, and the scale of increased marine sales will depend upon the geographic pattern of land-pit exhaustion.

required in relation to the restricted seaward access of many of the current landing ports (85% of current landing sites cannot take vessels of draught greater than 4.5m on neap tides, effectively restricting deliveries to vessels of 1500 cargo tonnes or smaller). The options open are:

(i) to close or relocate wharves, losing considerable geographical advantage in so doing.
(ii) to convert coasters to dredgers.
(iii) to build (or convert) barges and use sheltered water transfer systems (Section 5.1.3)
(iv) to build 1,500 tonne vessels.

To conclude, in maintaining marine landings at the levels implied in the Preferred Option there will need to be considerable investment in dredging plant within the next 5-10 years. Wharf relocation and rationalisation may help to simplify the dredger construction programme. Prospecting will need to be undertaken to ensure continuity of reserves towards the end of the period.

12.2.4 The possibilities for changed landing patterns 1986-2001

Given the major changes in licensed reserves and the dredging fleet required to ensure continuity of supply to the South East to 1996, it is difficult to comment on landing patterns beyond that date. The following general points can be made:

The fleet of large vessels serving the northern sector market will begin to need replacement soon after 1996.

With the RAG's 'philosophy' to increase the contribution from rail-borne and sea-borne sources of aggregates and in the light of continuing depletion of land-won reserves, increased use of marine materials can be expected, to meet up with imported supplies of crushed rock. Rail transhipment, and particularly the use of the Thames as a barge route for carrying aggregates into West London, may be considered more viable towards the end of the century (Sections 11.2.2 and 11.2.3) although the full potential of the M25 for lorry journeys has yet to be realised and evaluated.

Shortages of sharp (concreting) sand in West London and adjoining areas have been reported to the Consultants by the sand and gravel industry. If suitable supplies are not readily available locally (essentially for fine aggregate in concrete manufacture) then there may be some scope for dredging surplus sand from the East Coast banks, and elsewhere, and landing this at wharves along the River Thames.

In January 1986 agreement was reached between the French and British Governments to build a **Channel Tunnel**. Because of uncertainty over timing and aggregate demand, the DoE made no allowance for this project in their (latest) forecast of July 1981.

Construction is scheduled to take place between Autumn 1987 and Spring 1993. The total aggregate requirement for the UK works is estimated by the promoters at 2.8Mt, of which approximately 0.62Mt is for sharp sand and 1.2Mt is for gravel, the remainder being mainly crushed rock. In addition there will be a fill requirement of about 0.4Mt (R Ibell, CTG, pers. comm.). **The promoters see no reason why marine aggregates should not be used in this project provided, like other aggregates, they meet the specifications required and are available at competitive prices** (note; the concrete tunnel linings need not necessarily be constructed near the tunnel portal).

The ability to supply ports in France should also place UK marine operators in a position to provide part of the French aggregate requirement, estimated as 0.375Mt of sand and 0.725Mt of gravel.

The problems of supplying increased amounts of marine sand and gravel to the South East region, coupled with a continuing commitment by the Industry to supply Continental markets, accentuates the need for urgent releases of more offshore reserves. Problems relating to the release of known offshore reserves are seen as remaining a major constraint. Later, with continued depletion of reserves, the ability to locate new reserves could also present an additional limitation. Considerable emphasis should be placed upon optimum use of marine concreting aggregates and efficiency of dredging techniques in removing, and not wasting this increasingly valuable reserve.

12.3   EAST ANGLIA

12.3.1   The potential for marine aggregate landings

East Anglia comprises the counties of Norfolk, Suffolk and Cambridgeshire.

The **Preferred Option** for aggregates supply to the region took into account:

(i) a continued requirement for small amounts of exports to the South East region,
(ii) a low level of dependence on crushed rock,
(iii) severe doubts about the capacity of existing land-based reserves to meet the predicted high level of demand up to 1991,
(iv) that the continuing flow of planning permissions for sand and gravel was dependent to a considerable extent upon MAFF policies towards high quality agricultural land and its restoration following extraction. In the post 1991 period further supplies of sand and gravel might conflict with structure plan policies and the belief is that there will be increasing difficulties in maintaining supplies of locally won sand and gravel,

and the AWP recommended that:

(i) sand and gravel and small quantities of crushed rock would need to be imported from the East Midlands region,
(ii) in the light of revised DoE demand estimates it was not necessary to develop other options for sand and gravel supply but that
(iii) post 1991 it may be necessary to increase imports of crushed rock to replace previous indigenous sand and gravel supplies.

The **National Guidelines** indicate that the region will need to produce 6.2-11.2M tonnes of aggregate per year, 88% of which is from land-won sources, 8% from crushed rock and 4% from marine and alternative sources (0.3-0.5M tonnes per year). Some 9% of production will be for export, and a further 0.5-0.9M tonnes per year will require to be imported, both relating primarily to sand and gravel.

Marine landings up to 1983 were 0.2-0.3M tonnes per year. Subsequently landings have increased due to a one-off construction project, now complete; aggregate is landed at only one wharf.

There are no major **population** centres in the area, the 1.5M inhabitants being reasonably evenly spread throughout the region. Approximately 50% of the area lies within 50km of potential landing sites (Table 12.3, Figure 12.3).

Marine aggregate quality requirements are for concreting sand and gravel and building sand.

12.3.2   Constraints on marine supplies 1986-1996

The Preferred Option revised estimates indicated that an average landing rate of 0.3M tonnes per year could be expected, assuming negligible input from alternative sources. This figure has already been approached for the first half of the 1980's, and it is likely that two new wharf operations will open in Great Yarmouth and Lowestoft within

In this situation cargoes may be landed from further afield, such as from the Humber or Thames areas, although conserved East Coast reserves would ideally play the major supply role.

New wharf sites could be developed around the region's coasts, and existing or currently proposed facilities expanded. A barged supply to Norwich is being considered.

In the light of substantial increases in landings, investment in new ships dedicated to this market could occur, and would certainly be necessary by the end of the century when the existing dredgers reach the end of their working life.

The local proximity of the East Coast banks to the principal landing points in the region is seen by the Consultants as critical in providing a readily available alternative to land-won sand and gravel supplies. If towards the end of the period under consideration aggregates have to be dredged from the North Norfolk Banks/Humber area (because of demands on the local banks for London and the export market), the competitiveness of marine aggregates landed in the region may be lost.

12.4   EAST MIDLANDS

12.4.1   The potential for marine aggregate landings

The East Midlands region comprises the counties of Derbyshire, Nottinghamshire, Leicestershire, Northamptonshire and Lincolnshire.

The **Preferred Option** for aggregates supply within the region took into account:

(i) that the region is required to be a major exporter of land-won sand and gravel to the extent of 2.5-3.5M tonnes per year, primarily to Yorkshire and Humberside and the South East. Minor imports (0.55-0.72M tonnes per year) of 'sand and gravel and other materials' would be required to balance effects of geographical distribution of reserves.

(ii) there are large known reserves of sand and gravel in the coastal (Lincolnshire) zone.

(iii) apparent shortfalls to 1991 could be met by increasing production at existing sites, or releasing new reserves. Problems may arise in Northamptonshire towards 1991, and certainly in Northamptonshire and Leicestershire after 1991 increasing attention will have to be paid to the poorer quality higher glacial gravel deposits.

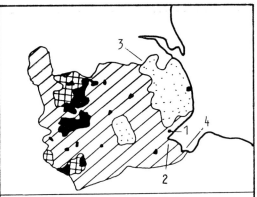

Numbers refer to Table 12.4

FIGURE 12.4   THE EAST MIDLANDS

1B BOSTON (T) 88mL, 4.5mD
2 FOSDYKE BRIDGE (RIVER WELLAND)
3 GAINSBOROUGH (inland port on R.TRENT)
4 SUTTON BRIDGE (proposed development on RIVER NENE)

TABLE 12.4   POTENTIAL LANDING SITES IN THE EAST MIDLANDS.

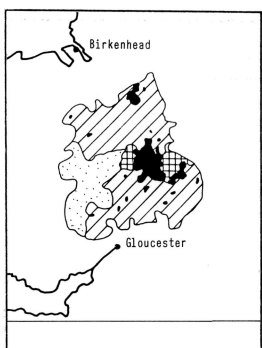

FIGURE 12.5 THE WEST MIDLANDS

The **National Guidelines** do not recognise any contribution from marine sources. Land-won sand and gravel production is forecast to lie between 8.6 and 12.6M tonnes per year to 1991.

Most of the area within 50km of the coast is sparsely populated (Figure 12.4), thus the **potential for marketing** marine aggregates from the few ports along the Wash coast (Table 12.4) is poor.

Only insignificant amounts of marine aggregate have previously been landed in the area (0.06M tonnes).

12.4.2   Conclusions

No marine landings are envisaged up to 1996. Barge transhipment from the Humber to the Trent Navigation may be considered in the late Nineties in relation to aggregate shortages, but the possibilities of such operations seem very remote.

12.5   WEST MIDLANDS

12.5.1   The potential for marine aggregate landings

This region, comprising the counties of Shropshire, Staffordshire, Hereford & Worcester, Warwickshire and West Midlands County, has no coastline.

In defining the **Preferred Option,** account was taken of the possibility of supply of marine dredged sand and gravel, but the AWP concluded that it would not be regionally significant.

The **National Guidelines** indicate that the region will need to produce about 197M tonnes of aggregate during the period 1981-1991, of which 52% (102M tonnes) will be land-won sand and gravel. Of the latter some 15M tonnes (8% of production) is required for export to other regions, principally the North West.

The AWP accepted that a continuing flow of permissions would be required to maintain production levels to 1991 and beyond. There may be problems in finding sufficient land in the south of the region to maintain production of sand and gravel, and consequently some pressure might be put on neighbouring regions. Further studies were needed in this respect.

At present very little marine aggregate reaches the region, mainly as special sands (DoE,1986). This results from

(i) accessibility; to the south the nearest port landing aggregates is Newport, a distance of 34km (although Gloucester, a potential landing site, is only 16km from the boundary - Figure 12.5). To the north the nearest port is Birkenhead, 51km distant.
(ii) only sand, not gravel, is currently available from the closest wharves.

12.5.2 Conclusions

There is unlikely to be any significant penetration of marine material into the region prior to 1996. To the year 2001 the only possible scope for development relates to hypothetical shortfalls of sand and gravel in the south of the region. The make-up of concreting aggregates from indigenous crushed rock and marine sands imported via Gloucester may be considered as a solution to this problem.

12.6 SOUTH WEST

12.6.1 The potential for marine aggregate landings

The South West region comprises the counties of Cornwall, Devon, Somerset, Dorset, Wiltshire, Avon and Gloucestershire.

The **Preferred Option** for aggregates supply within the region took into account:

(i) generally adequate permitted land-based reserves in the region to meet predicted demand,
(ii) a predicted gravel shortfall in the east Dorset production zone due to exhaustion of indigenous materials,
(iii) a probable small shortfall for sandstone (crushed rock) in the north Avon production area,
(iv) an anticipated increase in crushed rock demand in South East England,

and the AWP recommeded:

(i) increased imports of road surfacing aggregates,
(ii) increased imports of sand and gravel from Hampshire into the east Dorset zone if traditional supplies could not be maintained,

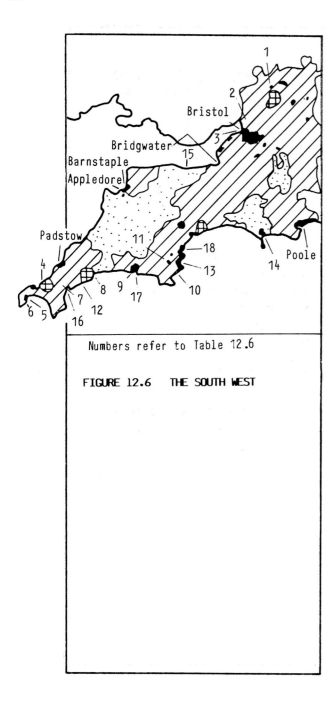

FIGURE 12.6   THE SOUTH WEST

(iii) increased export of crushed rock.

The **National Guidelines** indicate that between 1981 and 1991 only 2% of the region's aggregates needs would come from the sea, amounting to some 7-9M tonnes over the period, or annual rates of 0.6 to 0.8M tonnes per year between 1981-1991. In 1983 approximately 0.73 tonnes of marine aggregate were landed in the region. Marine landings to the region peaked in the early Seventies, at about 1.1M tonnes per year.

In terms of **accessibility** the marketing potential for marine aggregates within this region is good, there being few areas further than 50km from existing or possible landing sites (Table 12.6). Major **population centres** are also coastal (Figure 12.6), viz Bristol (population 440k), Bournemouth/Poole (280k), Plymouth (200k), Torbay (100k) and Gloucester (70k). Bristol's two wharves currently supply approximately 50% of the region's marine input, and the Poole wharf about 30%. The remaining four small wharves supply rural/small town areas along the southern shores of the Bristol Channel. A small percentage of total landings (<3%) is supplied from estuary dredging.

Aggregate quality requirements vary between the Dorset/Hampshire basin and the remainder of the area. In the former, marine sand and gravel supplies compete with similar land-won materials. In all other areas only building sand or concreting sand are landed, complementing existing land-won coarse aggregate supplies.

12.6.2   Constraints on marine supplies 1986-1996

Bristol Channel reserves are almost exclusively sand (building and concreting). The 40M tonnes of identified reserves also supply South Wales. Three companies each hold several licences in the Channel, giving a good basis for continuity of supply. At predicted rates of dredging sixteen years extraction for the South West market will only deplete the reserves by about one fifth (Note; reserves are also used by South Wales (Section 12.10.2).

The **Bristol Channel fleet** supplying the South West (normally five vessels) is old (age 17-23 years plus one vessels at 12 years). Although vessels may work on slightly beyond twenty-five years in this more sheltered environment, considerable reinvestment in ships will have to take place in the next 5-10 years. As Bristol can accept dredgers up to 1500 tonnes DWT, the availability of secondhand vessels might enable continuity of operation. The replacement of smaller vessels supplying the other Bristol Channel wharves poses more of a problem, and may require coaster conversion or offshore cargo transfer to barges (Section 5.1.3). **Wharf development or relocation** within the Avonmouth dock complex may also increase the size of vessel that can serve the Avon market.

The Bristol Channel fleet is currently underutilised, possibly by a factor of 25%.

South Coast reserves, particularly off the Isle of Wight, supply the Poole wharf. Only one company lands this material, thus the continuity of reserves cannot be commented on. The general point can be made that substantial reserves have been identified along the South Coast, and geologically there is good reason to believe that similar deposits lie along the coast in slightly deeper waters than those currently exploited, but available to the next generation of dredgers.

The present Poole operation can accept vessels up to 1500 tonnes cargo capacity, and there are no envisaged problems regarding vessel availability for the next ten years. Subsequently there may be scope for the construction of deep water berths outside the present dock area, should demand require it.

With the exception of the future provision of small vessels in the Bristol Channel, there appear to be no problems in meeting the supply of marine aggregate to the South West required in the Preferred Option. In the Bristol Channel identified reserves are adequate and projected increased demand can be accommodated by optimum utilisation of the present capacity of the fleet and shore-plant. Implications for the South Coast have been discussed in Section 12.2.

12.6.3   The possibilities for changed landing patterns 1986-2001

Meeting predicted indigenous material shortfalls with marine materials.

(i) Localised shortages of indigenous materials in Devon and Cornwall have prompted suggestions that minor operations could be set up at several wharves; future small vessel availability may preclude such developments unless offshore barge transfer was considered viable.

(ii) For the requirement for gravel in the north Avon area to be met by marine production, gravel reserves would need to be found within the Bristol Channel. None have been identified as yet, although there are suggestions that localised thick valley gravel deposits may exist (Section 3.9.2).

(iii) The predicted shortage of indigenous materials in the east Dorset production zone could be readily supplied from offshore via Poole. Firm planning

---

1A	GLOUCESTER (L)	9.5mB, 3.5mD
2A	SHARPNESS (L)	16mB, 6.5mD
3	AVONMOUTH (L)	
4A	HAYLE (T)	3.5mD
5A/D	PENZANCE (T)	2.8mD
6A	NEWLYN	6mD
7A	TRURO (T)	60mL
8A	CHARLESTOWN (L)	56mL, 9.7mB, 4mD
9A/D	PLYMOUTH - CATTEWATER	
10A	DARTMOUTH - study for 5000 GRT berth.	
11	TOTNES (T)	3.6m
12B/D	FOWEY	
13B	BRIXHAM (T)	90mL, 5.5mD
14B	WEYMOUTH (T)	45mL, 3.5mD
15D	WATCHET (T)	
16D	FALMOUTH	
17D	PLYMOUTH - MILLBAY	
18D	TEIGNMOUTH (T)	

TABLE 12.6 POTENTIAL LANDING SITES IN THE SOUTH WEST.

decisions would be required to enable the necessary investment in new landing sites to be undertaken (see Case Study No.12).

New marketing ventures are foreseeable for

(i) Plymouth; via Millbay or Cattewater docks, supplied from the South Coast.
(ii) Torbay; possibly via the projected Dart deep water terminal, supplied from the South Coast.
(iii) Gloucester area; via Sharpness and Gloucester docks, possibly related to future barging systems and offshore cargo transfer in the Bristol Channel.

Changed landing patterns pertain primarily to the South Coast of the region, supplied from South Coast reserves. Impact on the latter and the South Coast fleet have been discussed in Section 12.2

12.7 NORTH WEST

12.7.1 The potential for marine aggregate landings

The North West region comprises the counties of Lancashire, Cheshire and Merseyside and Greater Manchester (Figure 12.7).

In the **Preferred Option** it was recognised that:

(i) the supply of aggregate within the region depended on the continued import from adjacent regions; in 1977 this represented 48% of its requirements,
(ii) that in relation to local production, permitted crushed rock reserves would become exhausted around 1991, depending on the prevalence of the high or low predicted demand levels,
(iii) for indigenous permitted sand and gravel deposits there would be a shortfall of at least 16M tonnes relative to the required intra-regional production,
(iv) potential reserves of rock and sand and gravel had been identified within the region,

and the AWP recommended:

(i) over the period 1981-1991 some 13-17M tonnes of sand and gravel and 64-98M tonnes of crushed rock were required to be imported from adjacent regions,
(ii) potential reserves of rock could be released to comfortably cover the possible shortfall in regionally produced crushed rock,

(iii) potential reserves of indigenous sand and gravel could be released to cover the shortfall of this material at the low forecast level, although shortages of 6.25M tonnes would persist at the high forecast level.

The **National Guidelines** indicate that 37% of aggregate production within the region should come from land and marine sand and gravel sources, representing 33-50M tonnes over the period 1981-1991.

Although marine contributions were not specifically identified for reasons of confidentiality, this supply was implicit in the exercise; since 1975 there has been a steady supply of about 0.6M tonnes per year concreting and building sand into the region from offshore grounds. This figure represents approximately 14% of the region's consumption of sand and gravel, complementing 57% produced from local pit and beach extraction operations and 29% imported material.

Landings to Merseyside (presently about 0.4M tonnes per year) have dropped substantially since the early Seventies (when production peaked at 1.4M tonnes per year). This can be partly explained in terms of the effects of the development of the crushed rock market; **previously large quantities of gravel were dredged from Liverpool Bay, but the supply could not compete cost effectively with crushed rock.** This marketing conflict cannot be responsible for the whole of the fall however. As marine aggregates are thought to supply a high percentage of the Merseyside market today, the slump in landings presumably reflects the extent of the recession in this sub-region.

Aggregate is currently landed at four wharves (Figure 12.7), which in terms **accessibility** cover most of the western two thirds of the region within their 50km radii, an area with a poulation roughly estimated at 4M. However demand within the Cheshire area is primarily met by substantial indigenous reserves, and that of much of the Greater Manchester conurbation is met by sand and gravel imports from the North Wales and East Midlands area.

Aggregate quality requirements of marine materials are primarily for concreting sand and building sand, to complement crushed rock production.

12.7.2   Constraints on marine supplies 1986-1996

Reserves in Liverpool Bay are 10M tonnes of sand and 0.5M tonnes of gravel; the release of a further 10M tonnes of aggregate is awaiting CEC decision. Little of the Irish Sea has been prospected in detail, and there is a good chance of more extensive reserves being identified in the area, although probably further from the Merseyside market than the present three grounds (Section 3.2.7).

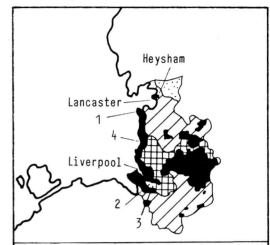

Numbers refer to Table 12.7

FIGURE 12.7   THE NORTH WEST

1A FLEETWOOD (L) 13.5mB, 4.8mD
2A GARSTON (L) 152mL, 19mB
3A MANCHESTER
    SHIP CANAL (L) 183mL, 24mB, 8.1mD
4D PRESTON (T) To fit in with
                yachting use.

TABLE 12.7
    POTENTIAL MARINE AGGREGATE
    LANDING SITES IN THE NORTH WEST.

The dredging operation in Liverpool Bay is worked as a consortium, and therefore the continuity of reserves can be confidently classed as adequate for the period to 1996 at present levels of supply. Should the supply to the combined North West and North Wales areas rise to exceed 1.0M tonnes per year however, the release of new reserves would be necessary.

Two vessels (occasionally three) work this area. The smallest of these (850 cargo tonnes) was built in 1961, and will probably need to be replaced within the next five years. The other vessels are fifteen years old. Replacement of a vessel of this size may prove a problem (Section 4.2.2), possibly necessitating the conversion of a coaster. A larger vessel could easily be run into the major landing sites in Liverpool and Birkenhead docks, which can accommodate ocean-going ships, but would not be economical at the present level of demand (the company already has a 2000 cargo tonne vessel laid up). The vessels serving the North West region are assessed to be 40% underutilised. The present level of landings could be maintained by the single vessel, with, as currently oocurs, occasional assistance from a visiting dredger. This situation however would not be very satisfactory from the point of view of interruption of supply resulting from breakdowns etc.

Wharf capacity is not a constraint. The existing Mersey wharves could be readily relocated to larger sites within the docks complex in the face of increased demand, and several port authorities with ample facilities have indicated willingness to accommodate landings; several of the latter were wharf sites for marine aggregates until the slump in demand of the early Seventies.

The North West region remains badly affected by the recession in the construction industry, and firms await an upturn in the market before deciding on future initiatives. The wharf and vessel capacities are underutilised, and can meet a substantial market increase without reinvestment, the exception relating to the approaching retirement age of one of the vessels and the problem of her replacement with the market in its current state. Licensed reserves can supply sand to meet a demand of 1.0M tonnes per year up to 1996 (40% of the region's requirements). There is therefore no problem in supplying marine aggregate at the present rate (20% of requirements) up to that date, to fulfill the Preferred Option.

12.7.3   The possibilities for changed landing patterns 1986-2001

Against a background of gross regional shortages of aggregates, the potential for increasing marine landings would appear favourable. However, crushed rock dominates the market for coarse concreting aggregates, and for concreting sands there is keen market competition along the south east and south west borders of the region, where demand is highest. Environmental issues also do not appear generally to be a major problem in this part of the United Kingdom, and marketing patterns are allowed to develop free from over restrictive planning constraints.

In this situation the outlet for marine material is related primarily to overall market demand. Should however, mineral planning authorities in North Wales decide against the release of new land reserves to meet predicted shortfalls (Section 12.11.1), increased marine landings to the North West region could readily fulfill the resulting shortfall in imports.

In the previous section the continuity of marine supply has been considered adequate for up to a rate of 1.0M tonnes per year, and to the year 1996. Beyond this rate and year new reserves would need to be identified, and new vessels will be required. Neither constraint is thought to be serious, particularly if encouraged by improved market conditions.

12.8   YORKSHIRE AND HUMBERSIDE

12.8.1   The potential for marine aggregate landings

This region contains North, West and East Yorkshire and Humberside.

In adopting the **Preferred Option** for aggregates supply within the region, the AWP took into account that;

(i)   imports, particularly from the East Midlands, would continue to meet the same proportion of demand,
(ii) sufficient reserves were believed to be available within the region as a whole to meet the anticipated levels of demand, although this would require the release of presently unpermitted reserves.

The **National Guidelines** indicate that

(i)   26% (3.9-5.5M tonnes per year) of the region's production should be land-won sand and gravel, of which 5% of the total production (0.7-1.0M tonnes) would be exported principally to the Northern region. A contribution of 0.14M tonnes per year might be expected from marine sources,

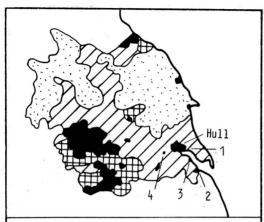

Numbers refer to Table 12.8

FIGURE 12.8 YORKSHIRE & HUMBERSIDE

1A HULL ABP. Deep water facilities
    (minimum 6mD)
2A GRIMSBY (L) 128mL,19mB,5.9mD
3A IMMINGHAM (L) 198mL,26mB,10.3mD
4D GOOLE

TABLE 12.8 POTENTIAL LANDING SITES IN THE YORKSHIRE & HUMBERSIDE REGION.

(ii) 1.9-2.7M tonnes per year of sand and gravel would need to be imported, the bulk of which would come from the East Midlands region.

Marine aggregate landings to the area have been steady at around 0.15-0.20M tonnes per year since the late Sixties; an increase in 1983 reflected the influence of a local 'one-off' project. An operaton was previously run at Immingham, but now there is only one landing site, at Hull.

In terms of **market accessibility** the potential for landing marine aggregates in this region is poor, as the main population centres are situated over 80km inland (Figure 12.8). The Hull area (population 350k) is the major market, where there is competition with road and barge transported aggregates from the East Midlands.

Aggregate quality requirement is for concreting sand and gravel and building sand.

12.8.2 Constraints on marine supplies 1986-1996

The region is supplied from the **reserves** off the River Humber; at 0.2M tonnes per year the impact on the 32M tonnes of licensed reserves is negligible.

The **vessel** currently supplying the market can be expected to be operational until 1996.

The **wharf site** currently used is constricted, and is possibly to be relocated; there is plenty of opportunity for new wharf sites at Hull (Table 12.8).

There is therefore no foreseeable problem in maintaining the present level of supply of marine-dredged sand and gravel to the Humberside area.

12.8.3 The potential for changed landing patterns 1986-2001

Near-coast areas. The recent construction of the Humber Bridge has improved market penetration from any new wharves set up along the Humber Estuary (Table 12.8), and may herald increased viability of landings in the future. Should non-release of land-based reserves (in either the East Midlands source or local areas) change the marketing pattern, wharf facilities are not thought to be a constraint. As with the supply to the Northern area (Section 12.9.3) substantially increased landings from the Humber grounds will require comparable releases of new offshore reserves, and capital investment in a dredger construction programme.

Inland areas. Although there are good motorway links between Humberside and the West Yorkshire conurbation, distances are thought too great for road transport. Barge transhipment using the Sheffield & South Yorkshire or Aire & Calder Navigation could be considered (Section 11.2.2); no comments can be made at this juncture on the future viability of such operations.

12.9 NORTHERN

12.9.1 The potential for marine aggregate landings

The Northern region comprises the counties of Cumbria, Northumberland, Tyne & Wear, Durham and Cleveland. Cumbria alone borders the Irish Sea, the remaining counties have coastlines along the North Sea.

The **Preferred Option** for aggregates supply within the region recognised that:

(i) rock reserves were sufficient to supply to regions's demands and allow necessary exports to the North West for many years, and certainly until 1991. The area most likely to suffer shortfalls if a high level of demand prevailed was Cumbria,

(ii) permitted reserves of sand and gravel in some parts of the region were insufficient to satisfy demand up to 1991, and that

(iii) the scope for new land-based reserves of sand and gravel within the region was extremely limited because of the generally inferior quality of the sand and gravel deposits in the region

and the AWP recommended:

(i) 12% of crushed rock produced should be for export to the North West,

(ii) sand and gravel and some crushed rock should be imported from Humberside and Yorkshire,

(iii) the balance of sand and gravel requirements should be sought from alternative materials produced within the region and from marine dredged sand and gravel.

In relation to sand and gravel requirements, the **National Guidelines** indicate that between 1981 and 1991 22-39M tonnes will be produced from land-won sources within the region, 8-13M tonnes will be imported and 10-13M tonnes should come from the sea, the latter representing 6% of the region's aggregates consumption, and an annual landing rate of 0.9-1.2M tonnes per year. The Guidelines also pointed out that any additional landings in SE England should not prejudice future supplies for the Northern region.

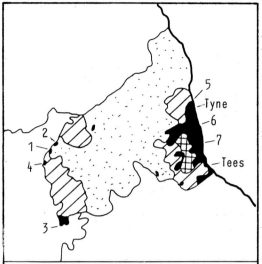

Numbers refer to Table 12.9

FIGURE 12.9   THE NORTHERN REGION

IRISH SEA COAST

1A WHITEHAVEN (L) 4.3mD
2B WORKINGTON (T) 75mL, 5.4mD
3B BARROW (L) Deep water facilities
4A SILLOTH (T) 80mL, 4.0mD

NORTH SEA COAST

5A BLYTH
6A SUNDERLAND (L) 142mL, 19mB, 7.2mD
7A SEAHAM (L) 90mL, 14mB, 5.0mD

TABLE 12.9   POTENTIAL LANDING SITES IN THE NORTHERN REGION.

Marine landings are not presently made along the Irish Sea coast, where indigenous materials satisfy demand. Landings along the North Sea coast began in 1969, increased to a peak of approximately 1M tonnes in the late Seventies, and are now at a level of 0.6M tonnes per year (note: includes estuary dredging in the River Tees, accounting for about 4-9% of the marine landings in the region, an operation which commenced prior to 1967). Three wharves are currently operated.

In terms of **accessibility** the marketing potential for marine aggregates is good, the main areas of high population lying within 50km of existing or potential port sites (Figure 12.9; Table 12.9); in particular the conurbations of Tyne, Wear and Teeside, with a combined population in excess of 1M persons.

Marine aggregate quality requirements are primarily for concreting sands and gravels.

12.9.2   Constraints on marine supplies 1986-1996

The North Sea coast of the Northern area is supplied principally from the Humber dredging grounds, although some material was brought up from the East Coast and Thames areas between 1974 and 1977. **Reserves in the Humber area** amount to 32M tonnes of concreting aggregates, with a similar quantity again prospected but refused due to fisheries conflict. Two companies currently deliver aggregate to the region, and a third may commence shortly. Each company has at least one licensed ground off the Humber. At the maximum level of landings envisaged within the National Guidelines, 1.2M tonnes per year, only 40% of the reserve will be utilised in the ten years up to 1996. This availability of reserves is not threatened by the present level of supply from the Humber to other regions of Britain, or to the Continent.

Two vessels currently serve the Northern market, and a third is soon likely to join this fleet. None of the vessels run exclusively to the NE coast, delivering also to Hull, the South East and the Continent. All the vessels are self-discharge, and will be less than twenty-five years old in 1996. At least two of the three companies currently involved have significantly underutilised vessel capacity.

Wharf sites are not a constraint on the NE coast, there being ample capacity on the Tyne and Tees, and also in the smaller ports such as Blyth, Sunderland and Seaham.

In conclusion, there appear to be no foreseeable problems in meeting the supply of marine aggregate to the east coast of the Northern region as required in the Preferred Option.

12.9.3   The possibilities for changed landing patterns 1986-2001

Meeting predicted indigenous material shortfalls with marine materials.

(i) The AWP has pointed out that if the contribution to the Preferred Option from either North Yorkshire/Humberside or marine landings were reduced, it would be necessary to consider whether additional supplies could be obtained from the other of these two sources, or from within the region.

(ii) The Working Party also indicated that existing permitted reserves would be largely exhausted after 1991, and that new reserves would have to be released.

Having regard to the apparently poor quality of remaining indigenous sand and gravel supplies, the possibility of future increase in marine landings seems high. Economies associated with the larger scale of such future operations should increase competitiveness with imported land-won aggregates.

The increase of landings on a significant scale on the NE coast would require:

(i) investment in new 5000 cargo tonne vessel(s) after 1996, as important elements of the fleet aged. Prior to this date a 30-50% increase in trade to the NE may be absorbed by currently laid-up or underutilised vessels,

(ii) extension of wharf facilities, which does not appear to pose problems,

(iii) the location of new or release of known reserves along the Humber/NE coast. The chances of finding good reserves closer to the NE ports seem slim, although one or two likely areas remain to be prospected in detail. The Humber reserves will be used primarily for the NE coast market in the future (see Sections 12.4 and 12.8 on local Humber demand), and at the maximum predicted rate of depletion within the National Guidelines existing reserves will be exhausted by 2001. Therefore any significant increase in landings to the Northern region will have to be matched by comparable release of reserves.

The future landing of aggregates on the Cumbrian coast is dependent upon the longevity of local land-based sand and gravel deposits, about which insufficient information is available; a potential problem has been identified in relation to the supply of crushed rock after 1991, which to some extent might be alleviated by landings of gravel. Wharf sites may be available at Barrow-in-Furness or in the Whitehaven/Workington area (Table 12.9). Aggregate is unlikely to come from existing Liverpool Bay reserves, especially if a high stone content is required. Suitable new reserves may be located off the Dumfries and Galloway shoreline (Section 3.2.7).

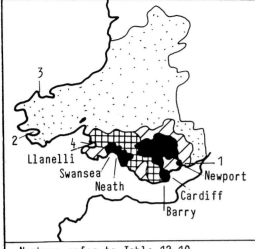

Numbers refer to Table 12.10

FIGURE 12.10   SOUTH WALES

1A	NEWPORT DOCKS	(L)
2B	MILFORD HAVEN	(T)
3D	FISHGUARD	
4D	LLANELLI NORTH DOCK	(T)

TABLE 12.10 POTENTIAL AGGREGATE LANDING SITES IN SOUTH WALES.

12.10   SOUTH WALES

12.10.1   The potential for marine aggregate landings

The South Wales region comprises Gwent, Mid South and West Glamorgan, Dyfed and Powys. Aggregate demand arises primarily in the industrial south east of the region (Figure 12.10).

The **Preferred Option** for aggregate supply within the region took into account that:

(i)   land-won sand and gravel reserves within the region were low, and could be exhausted at some time between 1985 and 1989,

(ii) permitted reserves of rock were greatly in excess of that required to 1991, although uneven geographical distribution could lead to some problems of supply. In Gwent in particular reserves were likely to be exhausted by 1991

and the AWP recommended:

(i)   maintenance of existing supplies using indigenous crushed rock for coarse aggregate requirements,

(ii) continued use of marine sand for concreting fine aggregate and building purposes,

(iii) extended use of secondary aggregates from mining tips to meet shortfalls in land supply.

The **National Guidelines** indicate that between 1981 and 1991 11% of the region's aggregate might be expected to come from offshore, amounting to some 11-20M tonnes in total, thus annual levels of production were estimated at 1.45-1.8M tonnes per year from 1981-1991; in 1983 approximately 1.4M tonnes were landed at the seventeen wharves in South Wales. Several county authorities encourage the landing of marine aggregate.

In terms of **market accessibility** the potential for marine aggregates is high, all the major South Wales conurbation (population c.1.5M) lying within 50km of existing wharves.

The **nature of the aggregate required** is primarily concreting sand, to complement quarried crushed rock supplies.

12.10.2   Constraints on marine supplies 1986-1996

The reserves supplying the South Wales region are derived from the Bristol Channel, where 40M tonnes of sand deposits have been identified. Five companies dredge several licences, and many of the grounds are shared, giving confidence for continuity of supply. No shortage of reserves is predicted in relation to the maximum envisaged rate of dredging, although taking into account the requirements of the South West region, reserves may be approaching depletion in the years immediately following 2001. The relative contributions of concreting and building sand within the total reserves is also not clear.

Seven or eight vessels are normally dedicated to the South Wales market, and their dates of construction range between 1961-1973, with only one vessel younger than 1969. Their cargo carrying capabilities range from 250 to 1900 tonnes, with a total landing capacity of approximately 7500 tonnes per day. Though vessels may work on slightly beyond twenty-five years in the more sheltered environment of the Bristol Channel, considerable reinvestment in vessels will have to take place in the next 5-10 years. Most of the vessels engaged in this trade today are small, although this is not warranted by the port facilities; Swansea, Cardiff, Barry and Newport have wet docks always accessible to vessels of 7m draught, and the rivers Neath and Usk (at Newport), in general, can accept vessels of up to 5m draught (which however have to dry out over the low-water period). Purchase and redeployment of 1500-3000 cargo tonne vessels from other areas may to some extent be possible, although there will be an equal demand for vessels of this size on the South Coast (Section 12.2). In view of the economics of vessel construction, a smaller fleet of 4500 tonne vessels making more use of the deep water dock facilities may ultimately be relied on.

The fleet is calculated to be currently underutilised by about 40%. Data supplied by mineral planning authorities support this calculation in that existing shore-plant can probably accommodate a further 425,000 tonnes throughput per year.

It can therefore be concluded that there are no apparent constraints on the supply of marine sands from the Bristol Channel at levels required in the Preferred Option up to the year 2001. Increased demand up to the maximum predicted level can be accommodated by existing reserves and dredger and shore-plant facilities. The replacement of the existing, ageing fleet does not in the Consultants' view pose insurmountable problems.

12.10.3 The possibilities for changed landing pattern 1986-2001

Meeting shortfalls in indigenous gravel supply with marine materials. A very important constraint in this respect is the apparent lack of gravel reserves within the inner Bristol Channel. Localised thick valley-fill sequences, possibly containing extensive surface gravel layers, have been postulated for the area (Section 3.2.9) but have not yet been identified. Localised gravel neposits are known to be present in several areas of the outer Bristol Channel. If the replacement of much of the present fleet with larger vessels becomes a reality, then there may be more scope for exploiting gravel deposits off the Dyfed or north Cornish coast, although little is presently known of the extent of such resources. Gravel material which was won from the area could be preferentially landed at Newport to help towards projected crushed rock shortages in this area.

New operations have been suggested for Fishguard and Milford Haven (Pembroke Dock) but the low level of demand in this thinly populated area does not make such ventures likely. New wharf sites within existing aggregate landing ports are a distinct possibility, including use of the wet docks at Newport. Such developments would be necessary to accommodate an increased level of marine landings above that predicted by the National Guidelines, possibly related to the discovery of offshore gravel deposits.

12.11 NORTH WALES

12.11.1 The potential for marine aggregate landings

The North wales region, which is thinly populated (total inhabitants c.500k), comprises the counties of Gwynedd and Clwyd. The most densely populated areas are found on the Liverpool Bay and Dee coast of the latter (Figure 12.11).

The **Preferred Option** for aggregate production in the region recognised that:

(i) North Wales is required to export some 33% of its sand and gravel production to the North West region,
(ii) rock reserves were sufficient to meet local demands and permit required crushed rock exports,
(iii) permitted land sand and gravel reserves of 22M tonnes were not sufficient, and that a shortfall of 5.3M tonnes could occur up to the period 1991 at the high forecast level of demand,
(iv) possible local shortfalls in crushed rock supply could occur

and the AWP concluded that additional reserves of land-won sand and gravel would need to be released before 1991. The possibility of significant marine supply was not considered.

The **National Guidelines** indicate that 30% of the region's aggregate production would need to come from marine and land-won sand and gravel, a total of between 17 and 26M tonnes over the period 1981-1991. Some 11-18M tonnes of this supply was to meet internal demands, of which a contribution could be marine. If the predicted shortfall in sand and gravel were to be made up by marine landings, this would amount to 0.84M tonnes per year over the period 1986-1991. Currently <0.1M tonnes are landed annually in a single operation at Bangor (Port Penrhyn).

The areas of greatest demand in the Menai Straits area and north and east Clwyd are within 50km radius of existing or potential landing sites (Table 12.11). Port space is at a premium along this coast however.

The **aggregate types** required are concreting and building sands, to complement crushed rock supplies.

<u>12.11.2   Constraints on marine supplies in relation to changed landing patterns 1986-2001</u>

The major constraint is **availability of port space.** The most suitably located existing facilities are found along the shores of the Dee Estuary, for example at Mostyn where the port operators have plenty of land for shore-plant facilities and are willing to consider aggregate landings (Table 12.11). The present landings at Port Penrhyn are hampered by tidal conditions and lack of space.

Current reserves are also reasonably limited. Some 10M tonnes of sand and 0.57M tonnes of gravel are worked from two licences on a shared ground off the North Wales coast and on two small grounds inside the Mersey Estuary. Some 0.6M tonnes of sand are presently supplied to the North West region from these grounds, thus if a substantial increase to the North Wales area occured a reserve shortage could be expected inside ten years. A further 10M tonnes of reserves are known from adjacent areas of the North Wales coast, but are awaiting CEC licence decision. The potential for discovery of new sand and gravel deposits offshore is reasonable however (Section 3.2.7).

Two vessels currently supply the North West area, of 2000 tonnes combined tonnage. Both, one because of its small size and the other because of its self-discharge capabilities, are capable of working 12 hours cycles for most of the time. In the

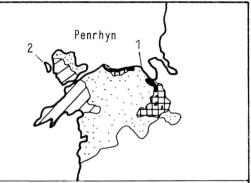

Numbers refer to Table 12.11

FIGURE 12.11   NORTH WALES

1B MOSTYN (T) 88mL, 5.4mD
2B HOLYHEAD 140mL, 4.6m maintained depth.

TABLE 12.11 POTENTIAL MARINE AGGREGATE LANDING SITES IN NORTH WALES

light of this capability the vessels are grossly underutilised (Section 12.7.2), and could possibly meet most of the demand of the North Wales region if there was no coincident increase in the North West markets. Furthermore the smaller of the vessels will be retired in the next five years and could be replaced with a larger capacity vessel; the same company has a 2000 cargo tonne vessel laid up. Vessel availability is therefore not seen as a serious constraint.

In conclusion, it is evident that although the National Guidelines do not look for a specific contribution to the North Wales region from marine landings, there is some scope for attempting to adjust the imbalance between projected land demand and supply through offshore resources. The availability of landing sites is seen as the major constraint, although reserve and vessel planning will need careful consideration. Shortfalls may be more realistically met by decreasing exports of land-won material to the North West region, and increasing marine supplies to the latter (Section 12.7).

A further possibility to ease supply problems in eastern Clwyd, already taking place to some extent, is the haulage of material from Birkenhead Dock (25km form the Clwyd border).

12.12 SCOTLAND

12.12.1 The potential for marine aggregate landings

Neither the National Planning Guidelines: Aggregate Working (1977) nor the National Planning Guidelines: Priorities for Development Planning (1981) provide any indication on the likely scale of aggregate demand from the country. The latest demand forecast figures for Scotland as a whole were published in 1980 by the DoE. Aggregates are not differentiated by type. For the period 1985 to 1991 an average of $24 \pm 4M$ tonnes per year is given. The 1977 Guidelines refers generally to the size of landbanks, indicating that planning authorities should consider whether adequate land to supply the demand of the market area already has planning permission. Also, that in basic zones (areas preferred for extraction) the landbank under planning permission ensures production for at least 10 years for each operator. Annual production of sand and gravel is currently around 10M tonnes.

Landings of marine aggregates have been almost nil; in 1980 3000 tonnes were dredged from Liverpool Bay and in 1982 2000 tonnes were dredged from the Thames Estuary. In addition, occasional 'one-off' fill jobs have occurred in Scottish waters such as Ayr Bay (2000 tones in 1984). Freshwater dredging in the River Tay provides limited quantities of sand and gravel.

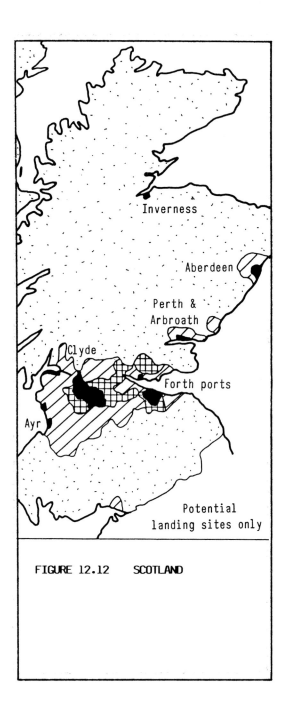

FIGURE 12.12 SCOTLAND

There would appear to be adequate land-won reserves of sand and gravel throughout the area. However considerable problems exist with the quality of indigenous deposits, notably shrinkage (Section 7.2.1), which has encouraged consideration of marine landings in recent years.

The adoption of the Scottish Development Department's policy by local planning authorities, to maintain at least a ten year landbank for each operator in their area, increases the difficulties of marine operators trying to establish markets in Scotland.

In terms of **accessibility** the marketing potential for marine aggregates within the Central Lowlands is good, most areas of high population lying within 50km of potential landing sites (Figure 12.12). Elsewhere in Scotland the population is too sparsely scattered to consider marine landings on any scale.

Aggregate requirements are for high quality concreting sands and gravels.

12.12.2 Constraints on marine landings 1986-2001

The major constraint on marine landings in Scotland **is the apparent lack of nearby offshore reserves.** Only limited reserves have yet been located in Scottish waters (Section 3.2.6). Suitable deposits may lie south of the Dumfries and Galloway coast in the Irish Sea (Section 3.2.7), but no prospecting has taken place in this area. Similarly deposits may occur in 40-50m depth off the Northumberland coast, which will be accessible to the next generation of dredgers. The identification of significant reserves in areas such as these could have a major impact on the Scottish market for marine concreting aggregates.

At present deliveries can only be contemplated from the Irish Sea or Humber reserves, over a steaming distance of some 250-350 nautical miles, which is presently not apparently viable on a regular basis.

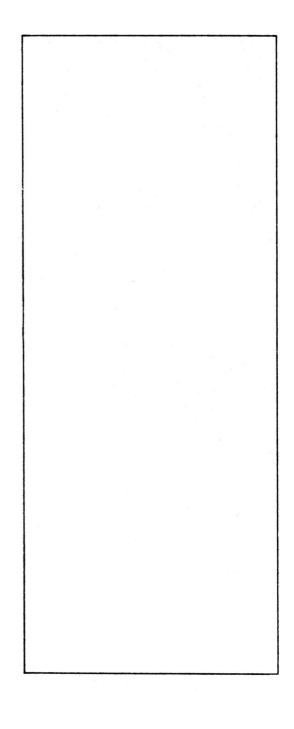

CONCLUSIONS AND RECOMMENDATIONS

CONCLUSIONS, RECOMMENDATIONS AND RESEARCH PROPOSALS

13

CONTENTS:

Summary of study objectives

Resource availability

Constraints imposed by conflicting offshore interests

Factors controlled by the industry and licensing authorities which could affect future developments

Locational and operational factors affecting onshore (wharf) development

Summary of research proposals

13.1　SUMMARY OF STUDY OBJECTIVES.

<u>13.1.1</u>　The National Guidelines (DoE Circular 21/82, Welsh Office Circular 30/82) form an important basis for the long-term planning of aggregates. As part of the national policy considerations the use of marine dredged aggregates is encouraged wherever this can be achieved without unacceptable damage to sea fisheries. Where there is a risk of coastal erosion licences to dredge are not and will not be granted.

<u>13.1.2</u>　The support by DoE for landings of marine aggregates is seen as reducing the major impact on the environment that land-won workings may cause, albeit a transitional use of land. However, the DoE recognise the need for a balance over the supply of construction materials from various sources because of possible conflicts of other interests with the exploitation of individual mineral resources. The importance of marine landings in saving unnecessary land releases is referred to in Section 4.1.1. As part of this balance between aggregates from the sea and from the land the Consultants see an increasingly important role for the mineral planning authorities (Section 13.4.1).

<u>13.1.3</u>　The marine aggregate dredging industry has responded to the emphasis given in the National Guidelines by investing in ships, wharves, prospecting and licences, such that the industry currently supplies some 12 Mt of sand and gravel per year to the UK in addition to some 3 Mt per year landed at ports on the Continent. This has been a very positive achievement and has given the UK one of the largest marine mining industries in the world, a full description of which has been undertaken as part of this study and is reported elsewhere (DoE, 1986).

<u>13.1.4</u>　**The main objectives of this study have been to isolate and examine the constraints which will affect the future of the Industry, and to recommend pathways that may be followed by both Central Government and the Industry to alleviate constraints and help encourage a healthy expansion of the sea-dredged aggregates trade.** The results of this investigation have been reported in the previous chapters, and relate to:

　i)　　Resource availability.

ii) The degree to which resources are constrained by other interests in the offshore environment.

iii) Factors largely controlled by the Industry and the licensing authorities which will affect future development. These include ways in which technology and management may advance to better utilise resources and resolve conflict with other 'users' of the offshore environment. Also included are appraisals of the ecomonics of offshore aggregate mining, the problems of marketing marine aggregates, and quality control and user prejudice constraints.

iv) The location and operation of onshore (wharf) sites, and the extent to which future growth will be controlled by such factors.

13.1.5 The Consultants' conclusions are ordered in the above manner. Recommendations (including research proposals) are made, where appropriate, adjacent to the relevant conclusion and are identified by **bold type.** Research proposals are also reiterated in the final section.

13.2   RESOURCE AVAILABILITY

13.2.1 The availability of reserves to meet projected market demands to the end of the century varies regionally. Three situations prevail:

i) Areas where there are essentially sufficient reserves already licensed. This applies to the Humber and Bristol Channel areas. Prospecting and Production Licence applications in these areas will continue however, both to ensure continuity of supply into the next century, and in readiness for changed market demand (in quantity or aggregate type).

ii) Areas where serious deficiencies have been projected and although new reserves have been identified, Production Licences have been largely refused or are undecided because of objections predominantly concerned with fisheries and coastline stability. The East Coast, Thames Estuary, (east) South Coast and to some degree the North West region fall into this category (note the repercussions for the SOUTH EAST REGION in particular). To an extent the problem may be solved by the discovery of further reserves, but resolution of outstanding issues that will release reserves already applied for is looked for in many instances by the Industry (Section 13.3).

iii) Areas where prospecting to date has not identified any large scale, suitably located reserves. This situation occurs in the North East, where existing markets are supplied via a long steam from the Humber area, and in Scotland, where potential markets in the Central Lowlands are as yet unsupplied from marine sources.

13.2.2 It is clear that in all areas prospecting remains a priority activity. Prospecting is an expensive operation. In Section 13.4.3 methods are reviewed whereby prospecting may be made more cost effective.

13.2.3 Nationally there is very poor awareness of where and in what quantity marine aggregate resources may exist outside established reserve areas. A considerable body of data from many sources exists however, describing the nature of our continental shelf deposits. Suitably analysed, and added to where necessary by new field data, this body of information could provide an invaluable REGIONAL RECONNAISSANCE SERVICE, for the direction of new prospecting ventures, and the planning of future marine aggregate resource exploitation. **It is recommended that such a data bank and advisory service be set up and funded, possibly within the British Geological Survey.**

13.2.4 The investigation has shown that there is no clear evidence for long-term continuity of reserves, and thus no basis for assuming that marine aggregates will be available, for example 50 years hence. This poor level of knowledge will improve if regional reconnaissance exercises are undertaken, as concluded above. **For the present however, there is a need to ensure the most efficient utilisation of reserves, avoiding partial dredging and over-sanding situations.** It has been concluded by the Consultants that there is only minimal awareness at the present time of the efficiency with which grounds are currently dredged. Future developments in prospecting and reserve management which should improve this situation are addressed in Section 13.4.3.

13.2.5 A substantial proportion of UK marine aggregate reserves are EXPORTED to the Continent (about 3 Mt per year, cf 12 Mt landed in the UK - see note in Section 12.2.2). The Consultants believe that the export trade is less remunerative than landings to the UK, but is necessary to maintain continuity of work-load for the presently underutilised fleet. The Government consider that it would not be in the nation's interest to restrict exports; they assist in the overseas balance of payments. Logistical considerations presently dictate which grounds are used for the export trade. The East Coast and Thames Estuary grounds are a major source of export material, a factor which may not be in the National interest considering existing and future demands in the South East and East Anglian regions, and the current problems of licensed reserve shortages on these grounds. The pressures are likely to always be present.

13.3 CONSTRAINTS IMPOSED BY CONFLICTING OFFSHORE INTERESTS

13.3.1 In comparison with the 189 Mt of reserves currently licensed, over the past twenty years some 87 Mt have been refused licence, and a further 70+ Mt are awaiting licensing decision. In certain key areas (notably the SOUTH EAST region, Section 13.2.1) the resource problem is serious, and there is considerable concern regarding the continuity of reserve availability before the turn of the century.

13.3.2 Of equal importance, it is apparent that the contentiousness of licence applications has increased with time, and a situation now prevails where several major applications have not been decided upon after many years of consultation. This trend is undoubtedly attributable to the increased awareness of environmental issues generally within the UK over the past 10-20 years. This apparent inability to resolve issues may suggest shortcomings in the present approach to licence determination.

13.3.3 Certain licence refusals are unavoidable, viz cable and oil/gas production safety zones, certain critical navigation zones, areas very close inshore (from the coastline stability viewpoint) and sensitive fishery areas (spawning and nursery grounds). Because of confidentiality problems, the Consultants were unable to ascertain what percentage of the total refusals fell into this category. The frequency of this type of objection can only worsen with time, due to the proliferation of cables, increased nearshore oil exploration, and the relinquishment (through exhaustion) of existing close-inshore dredging grounds which where licensed prior to the introduction of current coastline stability criteria. **In view of the increasing value placed upon aggregate supplies the Consultants recommend that CEC and government departments ensure the greatest consideration for offshore sand and gravel resources in their coordination of applications for exclusive uses of the sea bed.**

13.3.4 Some licence applications have been refused as a result of clear-cut decisions; 'safety' criteria were applied and sufficient data were available to show that conditions within the proposed dredging grounds would be problematical. However, it has been made clear to the Consultants that some decision making criteria have changed in the past ten years, perhaps in some instances only subtly but significantly enough for applications to be reviewed in a new light. Furthermore, this study has identified general areas where fundamental research is required to better future decision making (Section 13.3.5.). As all 'safety' criteria are imposed from the point of view of extreme caution, the effects of new criteria based upon increased scientific knowledge of the environment could have the effect of releasing previously refused licences. **It is therefore recommended that all 'proven' licence refusals are categorised according to the criteria they failed upon, and that as decision-making**

criteria are refined, all such past licence refusals are automatically reviewed. The imposition of this system would require the accurate definition and listing by the various bodies involved of all decision-making criteria, and an immediate review of all past refusals in relation to updated current licensing standards.

13.3.5 Possibly most licence applications that have been refused were unproven, ie there were insufficient data to allow a responsible decision to be made. To some extent poor knowledge may have related to the decision making criteria, as concluded above. **The Consultants recommend that the remedy is for basic research to be undertaken; important topics for such research have been suggested during this project, and are reviewed in Section 13.6.** In other instances the lack of data was site-specific, namely that reserve (prospecting) data or environmental (biological/sedimentological) data were inadequate. Such data can either be collected by the dredging companies during the prospecting period (Section 13.4.3) or by the consulted bodies (notably HR and MAFF) after the licence application is submitted. The latter approach has to date been relied upon by the Industry, and is largely responsible for the long delays between licence application and arrival at a decision. For the future the Consultants recommend that:

i) **CEC should require sufficient data on reserves and environment to be submitted with each licence application (Section 13.4.3). Data standards could be readily set on the basis of existing experience, and/or**

ii) **CEC should adopt a system of forward planning of reserve release, giving time for generalised environmental problems associated with quite broad areas to be considered prior to licence application (cf oil block release system).**

13.3.6 The absence of detailed monitoring of dredging activity has been suggested as playing a significant role in licence refusal. Whereas authorities responsible for coastline protection and fisheries can be accused of relying upon too vague boundaries to define prohibited dredging areas, any attempt to encourage tighter definition of sensitive areas are valueless in the absence of close control of dredger activity (Section 3.4.3).

13.3.7 Even with improved scientific input to licensing decisions, resolution of objections may remain problematic, due to factors such as:

i) The economic and sociological consequences of displacing fishing areas, although this impact may be lessened by the operation of closed dredging seasons, limiting dredging to certain parts of grounds at any one time, and care in leaving dredging areas in a condition where they can be used again as fishery grounds (Section 13.4.3).

ii) Prejudices and lack of co-operation resulting from insufficient knowledge relative to the problems involved and past 'misdemeanours' on the part of the dredging industry.

The Consultants recommend that a long-term improvement in these two situations needs to be achieved from public relations exercises initiated by the dredging industry, and improved contact and co-operation generally, and the introduction of electronic activity monitoring.

13.4 FACTORS CONTROLLED BY THE INDUSTRY AND LICENSING AUTHORITIES WHICH COULD AFFECT FUTURE DEVELOPMENT

13.4.1 Licensing

A major problem exists regarding the determination of licences as a result of conflicting environmental interests (Section 13.3). As has been suggested, the solution lies with improved scientific, economic and sociological understanding of the issues involved. To some extent however the informal nature of the present licensing procedure is not ideal, particularly as:

i) Consultation is confined to parties potentially affected by the proposed marine operations, and does not take regard of effects of alternative on-land development should the licence be refused,

ii) there is no overall time limit on the determination procedure, and

iii) there is no clear-cut system of appeal.

It understood that the licensing procedures are currently under review, and the Consultants recommend that consideration is given to the above points, bringing offshore reserve licensing more into line with the planning procedures that have been adopted on land.

In particular, the Consultants recommend that mineral planning authorities are more closely involved in licensing decisions, as they are in the best position to introduce an element of balance between the impact of onshore and offshore operations. The 12 M tonnes of marine aggregate landed annually is comparable to an onshore production from

about 240 hectares of quarry workings. In instances where the mineral planning authorities establish that a market requirement can be supplied from either onshore or offshore expansion of operations, MAFF can be asked to decide upon a preferred option in terms of agriculture or fisheries effects.

13.4.2   Economics

The primary economic factors affecting the existing operation of the marine sand and gravel industry are:

i) Road haulage costs: both land and marine operators are controlled principally by the proximity of source and supply; the trend is towards wharf proliferation in order to minimise transport. Thus distance from wharf to market is critical.

ii) The price of marine material (ex wharf) is in general more than land-won material (ex pit). Thus the competitive penetration of marine material is normally less than land-won. Three major variants on this model have been identified:

   a) Areas which lack indigenous land-won material.

   b) Areas of land subject to environmental pressures.

   c) Areas of poor quality land-won material (primarily North East England and Scotland), all of which encourage marine supply.

iii) Proximity of wharf and dredging ground has always been, and still is, the major factor controlling the price of marine material. This is largely influenced by fuel costs. With depleted inshore reserves and increasing environmental pressures, the trend is towards working more distant deposits on 24-36 hour cycles, with associated higher production costs.

The constraints imposed by these factors control the marketing of marine aggregates. Large concerns, with both land and marine operations, have in many areas been able to optimise dredger utilisation (husbanding onshore assets and making best use of depreciating offshore investments). This partially explains the increased percentage share by marine landings of the total sand and gravel market to 1980 and thereafter maintaining it at that level of around 12-13%.

Smaller companies have had to adopt a more aggressive approach to marketing and often there seems limited scope to increase landings overall because the market can only absorb a certain tonnage of material at a given time.

A major constraint to future operations has been identified by the marine sand and gravel industry in that investment in new, multi-million pound dredgers does not make commercial sense UNLESS adequate offshore reserves are available (Section 13.3). Although two orders for new dredgers where placed in 1985 by one company, the majority of companies would prefer to see a potential for return on invested capital before ordering new vessels.

The marine industry is placed in a difficult position over its forward planning: it has expressed its will to increase landings, where it is supported by Government and many local authorities, but inadequate licensed reserves are proving a major obstacle to investment. Without sufficient reserves the Industry is also reluctant to seek new wharf locations (apart from in rationalisation exercises).

This study has confirmed that major investment in vessels will be required during the next ten years in the following regions:

i) The South Coast of the South East region,

ii) the Bristol Channel in the South West and South Wales regions, and

iii) Liverpool Bay in the North West region.

13.4.3 Technology

Vessel characteristics: little development is expected in this field within the period under consideration. The Industry appears to have identified an optimum vessel size of about 4500 GRT, and any new vessels built may be expected to be of this size. Considerable vessel replacement will be necessary during the 1990's. With the demise of older small dredgers, and possibly the inshore shallow dredging areas, it is not clear how existing small wharves will be supplied; one solution of which the Industry has some experience may be sheltered water transfer to barges.

Dredging plant: the next generation of dredgers may be expected to use in-pipe dredge pumps, thus enabling dredging to about 45m depth. This will enable better utilisation of existing deeper water licensed reserves, and expansion into prospecting new areas further offshore.

Seabed sensing and monitoring techniques: the Industry currently makes little use of the most advanced techniques for navigation and underwater sensing. Using existing technology advances could be made in accuracy of ship positioning, data handling and storage, seabed surface mapping, and environmental data collection. Although the Industry generally may tend to resist such a move on cost grounds, in the Consultants' opinion there are several (financial) benefits which would accrue:

i) More accurate mining practices combined with better prospecting data would enable pre-planning of the final condition of the seabed. Thus coastal stability and fishery objections would be less generalised and more likely to favour dredging (Section 13.3.5).

ii) More accurate mining practices combined with better prospecting data should enable optimum recovery of aggregate content of a reserve (Section 13.2.4) and possibly prevent situations such as over-sanding. Avoidance of serious cargo contamination and prediction of cargo quality should also be more feasible.

iii) The generation of environmental data, the adoption of activity recording systems and progress monitoring of bathymetry would be invaluable in resolving licence application issues and would provide information for public relations exercises throughout the lifetime of the mining project to allay any unjustified fears regarding the effects of dredging.

Beyond the adoption of existing equipment for underwater measurements, three areas of underwater sensing have been identified where development is needed, and where the offshore aggregate industry could be expected to initiate or collaborate with research effort. Improved techniques for geophysical and geochemical profiling of seabed deposits could increase the cost efficiency of prospecting. Devices for swifter and more efficient retrieval of cores from gravel substrates would overcome most of the existing problems of insufficient information and the precise nature of deposits in a licensed reserve. And finally, systems for the in-pipe sensing of cargo quality could eliminate cargo contamination problems, and enhance a captain's ability to respond to day-by-day quality requirements of the wharves to which he is delivering. **A review of development requirements in these fields is recommended in the first instance.**

13.4.4 Quality control and user prejudice

The problems associated with aggregate quality are not confined to marine sand and gravel. The British Standards Institution have published a number of Standards relating to the quality control of aggregates and their use in concrete. Further Standards are in the course of preparation. Of particular concern to the Industry are chlorides and alkali-silica reactions in concrete. The marine industry has shown that with adequate processing of their materials, the use of marine dredged aggregates is no less inferior to sand and gravel excavated from the land.

Many wharves, although not all, experience problems over sales primarily due to customers' prejudice to the use of aggregates dredged from the seabed. The Consultants suspect that many building contractors and their architects/engineering advisors are recommending avoidance of marine aggregates largely through ignorance of the inherent properties of marine sand and gravel.

Greater awareness is necessary by specifiers and customers alike that marine dredged aggregates can be used more widely. The publication of various results of research into building materials and the imposition of tighter, more exacting, aggregate testing standards, should enable greater use of marine aggregates.

The Consultants recommend that **the sand and gravel industry should collectively address itself to combatting user prejudice. Publicity should be made in trade and professional journals used by architects, engineers, surveyors and building contractors explaining the uses and extent to which marine aggregates are already used without ill effects.**

The Industry may wish to consider what other measures could be taken to allay fears, eg whether or not continuous monitoring of chlorides, as opposed to spot sampling, might satisfy some customers.

Reference has been made to differences in the classification of GRAIN SIZES of sediments, with the aggregates industry following BS 882. Outside the aggregates industry, but nevertheless associated with descriptions of grain size, engineers and geologists use other scales with different limits for the division between coarse and fine aggregates and different methods of grade description.

It is suggested that the BSI be asked to examine the possibility of producing a combined Standard, removing unnecessary anomalies where possible. Clarification on terminology is seen as important to avoid misunderstanding, and to promote better use of aggregate prospecting data in combination with data generated by other offshore research.

13.5 LOCATIONAL AND OPERATIONAL FACTORS AFFECTING ONSHORE (WHARF) DEVELOPMENT

13.5.1 Wharf availability as a constraint

Around Great Britain generally there are no major problems associated with wharf location which will affect the prospects of maintaining or increasing landings of marine aggregates to the various regions. The single exception to this is the North Wales region. The West Midlands (which is land-locked) is constrained by distance to the nearest wharves.

This study has revealed that many ports around the UK coast currently not receiving aggregates could be utilised in initiating landings, and that there is scope for improving and enlarging wharfage facilities in most of the existing landing ports.

As happens at present, for a variety of reasons, some established wharves may close but this does not imply that the tonnage landed will decrease in the area overall. Often companies relocating wharves will consider rationalising existing wharves, especially if there are operational difficulties such as security of tenure, small congested sites and access problems.

13.5.2 Inland wharf location

Landings of marine aggregates offer in some instances ideal conditions for alleviating pressures on road transport as a result of inland penetration along the waterways. This may either take the form of dredgers steaming long distances up large estuaries, or of transhipment into inland waterway barges.

Grant aid is currently used by the sand and gravel industry for the movement of materials by rail only. Although the Transport Act 1981 provides for grant to cover the movement of materials along inland waterways, its application is currently very limited. This is because no operator at present moves marine sand and gravel for onward transhipment by barge, despite the fact that virtually all existing marine sand and gravel wharf locations fall within the 'distance of opposite banks' criterion for an inland waterway.

Since the main purpose of Section 36 is to provide grant aid for environmental reasons (primarily to remove lorry traffic from roads), the Consultants recommend that it would be appropriate for this Section of the Act to be extended to include existing and proposed wharves <u>where river traffic can be increased, at the expense of road haulage.</u>

It appears to the Consultants that the only satisfactory case that could be made out for grant aid for marine aggregate shipping within the context of Section 36, relates to the provision of specialised vessels which can navigate inland waterways. Three categories of vessel can be envisaged:

i) Sea-going vessels of special design/dimensions enabling navigation of inland waterways (eg existing vessels plying to Central London).

ii) Barges designed to ply between small ports and sheltered-water transfer sites (Section 13.4.3).

iii) Ordinary inland waterway barges.

The Consultants recommend that vessels, sea-going or otherwise, that are specifically designed to pass beneath bridges and/or designed with shallow draughts should be eligible for a Section 36 grant. The majority of aggregate dredgers would possibly not be eligible: an existing example used by the Industry are the low-profile 'Bow' vessels where draught and height of superstructure are all important.

With regard to barging the Consultants recommend that **the Industry should investigate the practicalities and financial implications of joint ventures with established barging companies.**

13.5.3   Accessibility from the land

It is very likely that the majority of sand and gravel landings will continue to be sold ex-wharf using lorry transport in view of the proliferation and small size of wharves around the UK. Many of these wharves are not, and cannot be, physically connected to a rail network. In addition, many of these wharves are incapable of handling an increased throughput. Improvements to the trunk road network serving urban areas (eg M25) provide a greater incentive to retain lorry transport as the means for distribution.

The opportunities for increased use of rail transport are not only limited to the South East (where there is a guaranteed demand arising from within the London area) but also are unlikely to come about for many years.

13.5.4 Cargo rehandling

No major technological innovations are apparent. Only minor refinements are likely to occur as more emphasis is placed on dredger utilisation and turnround times (Section 4.2.2). These will take the form of greater use of self-discharge methods resulting in shorter unloading times, although there is often little scope for time and motion improvement in instances where the turnround time of a vessel is controlled by the tidal cycle and not the rate of discharge.

13.5.5 Materials processing: offshore and onshore

The problems of shipboard processing experienced in the past still apply today. There are no major advantages offered by processing at sea and in fact some long-term ecological problems might arise with additional solids in suspension. The trend towards simplified deck plant to clean up 'as-dredged' cargoes to approximate shore requirements will continue. There is apparently no future for other forms of onship processing.

With onshore processing equipment little change in the foreseeable future is likely. Small scale refinements to individual components will continue, particularly where any tightening of national quality controls are introduced.

Improved hydraulic classification might be an area for such development in separating sands of different grades and to ensure minimum loss of products through silt disposal.

Wharves located outside the main industrial areas ie in rural or residential locations, might see greater use of low-profile processing plant both for washing/grading and value-added plant (sometimes referred to as semi-transportable plant) to reduce the environmental impact of operations.

13.5.6 Environmental constraints

In general there is little conflict between the aggregate producers and their impact on the surrounding community since many wharves are located in established dock areas or are removed from areas of human habitation. Where conflicts do arise these can generate enormous local feeling but in the Consultants' view these are of limited significance only and do not merit any recommendations for further research.

Pressure may come from local authorities as run-down dock areas are progressively redeveloped. In these situations suitable alternative sites must be negotiated between mineral operator and planning authority to ensure no loss of markets and continuity of supplies. The view prevailing within local authorities involved with marine aggregate wharves is that they form part of the commercial life of an area, and often the passage of dredgers along a river is one of the few signs left to indicate the prosperity of a town or city which has previously seen considerably more river traffic.

The Consultants recommend that it may be instructive for a local authority to initiate investigations to gauge public opinion from residents and visitors to an area regarding the presence of dredgers plying to and fro through areas of concentrated water amenity usage.

Similarly, a study into the potential impact of the passage of dredgers, maintenance dredging of channels and the landing of cargoes within an SSSI or other areas of ecological importance could usefully contribute to future decision making both for new wharves and where increased landings are proposed. These studies could be undertaken by local authorities in co-operation with the wharf operator, yachting associations, the Nature Conservancy Council and/or local naturalist trusts.

13.6   SUMMARY OF RESEARCH PROPOSALS

13.6.1   Regional reconnaissance research

Initially requiring a desk-top programme to collate and analyse offshore data from various sources with the aim of identifying national marine aggregate resources, this research might expand to involve some data collection on a regional scale (Section 6.2.2).

13.6.2   Fundamental research relating to licensing criteria

As a first step the Consultants recommend that a review of the reasons for which past licences have been refused is undertaken, thus clarifying research priorities. The need for research on the following subjects is, in the Consultants' opinion, already clear:

Shoreline stability (Section 9.4)

i)   Development of models defining site-specific inshore dredging limits in relation to gravel mobility, water depth, wave climate and tidal regime.

ii) Field investigations to improve knowledge of the importance of active offshore sand supply to beaches at a range of potentially sensitive localities.

iii) A review of scientific methods currently used in dredging related research.

Environmental and Fishery (Section 10.5)

i) Field investigations of a series of exhausted dredging areas to establish residual effects on substrate, recolonisation models and the possible role as 'fish reserves' of zones not reusable by the fishing industry.

ii) Studies of the impact of dredging on herring and crustacea breeding.

iii) Establishment of the long-term economic value of inshore fishing areas for comparison with returns expected from dredging.

13.6.3 Research and development into new techiques for prospecting and reserve management

Three areas of potential development have been identified:

i) Geophysical and geochemical techniques (Section 6.2.3)

ii) Deposit coring systems (Section 6.2.4)

iii) In-pipe aggregate quality monitoring (Section 5.3.1)

These requirements are at least partly shared by the offshore civil engineering industry, and the possibilities of co-operative research and development should be looked into.

13.6.4 Environmental impact of wharves

Only local studies are seen as necessary where potential or actual conflicts arise between the passage of dredgers and/or wharf operations (Section 11.4). Local authorities should initiate such investigations as and when necessary to assess the degree to which the onshore activities (including inshore shipping movements) of the marine aggregates industry enhance or detract from the amenities of an area.

13.6.5   Site-specific investigations and routine monitoring

Although this type of investigation does not strictly come under the heading of research, it should be noted that **the Consultants recommend the Industry to make an increased commitment to generation of prospecting and environmental data, on a routine basis, with its associated costs.**

13.6.6   Research funding

Three principal sources of research, development and monitoring funds can be identified:

i) Direct funding by individual companies, which it would seem reasonable to largely direct towards site-specific licence application and routine monitoring studies.

ii) The setting aside of a fraction of the CEC royalty as an 'improved management fund'. These monies might initially be directed towards the implementation of pilot schemes to encourage the Industry into better reserve management practices, and associated research and development. Such a fund could also be used to partially reimburse prospecting costs in instances of licence refusal.

iii) Central Government (DoE, MAFF) financed research, which would possibly be best utilised in more fundamental, 'one off' research programmes, as identified in Section 13.6.2.

Research projects are required to be cost effective. With a few exceptions the topics that have been suggested are intended to provide the minimum requirement to place licensing decisions on a sound scientific basis. Without this basis the Consultants believe that satisfactory resolution of licence applications will continue to prove difficult or even impossible. In this light, the return on funds invested in research is seen as the long-term survival of the offshore aggregate dredging industry.

APPENDICES

APPENDIX 1

GLOSSARY OF TERMS

Admixture (or Additive)   A material other than coarse or fine aggregate, cement or water added in small quantities during the mixing of concrete to produce some desired modification in one or more of its properties.

Ambient	That which encompasses.
Anoxic	Deficient in oxygen.
AWP	Aggregate Working Party
BACMI	British Aggregate Construction Materials Industries
Bathymetry	The science of sounding seas and lakes.
BCI	Blue Circle Industries
Bed-load	Sediment transported in contact with the bed.
Benthos	Flora and fauna of the sea bottom.
BGS	British Geological Survey (formerly IGS)
Biosphere	The part of the earth and its atmosphere in which living things are found.
Biota	Flora and fauna.
BRE	Building Research Establishment
BSI	British Standards Institution
Carbonation	Particularly important for reinforced concrete. CO_2 in the air and water react to form $Ca(OH)_2$ which carbonates concrete towards embedded metal. This reduces the alkalinity of concrete and its strength.
CEC	Crown Estate Commissioners
Chemosensory	Sensitive to chemical stimulus.

Cretaceous	Geological period, covering 136 to 65 M years ago, representing last part of the Mesozoic era.
CTG	Channel Tunnel Group
DAFS	Department of Agriculture and Fisheries for Scotland
Demersal	Found on or near the bottom of the sea.
Detritus	A mass of substance gradually worn off solid bodies: an aggregate of loosened fragments, especially of rock.
Devensian	Last geological stage of the Pleistocene epoch, covering 50,000 to 10,000 years ago, representing the last main advance of glaciers.
Discrete	Separate, consisting of distinct parts.
DoE	Department of the Environment
Dolphin	permanent moorings for vessels set in bed of river with steel or wooden piling and strengthened against vessel impact by some form of binding or fixing the their heads.
DTp	Department of Transport
DTI	Department of Trade and Industry
Echo-sounder	Depth measuring device using sound waves.
Ecosystem	Assemblage of flora and fauna and the interrelationships of their life functions.
Epibenthos	That section of the benthos living on the sediment/rock surface.
Epifauna	Animals that live on the surface of a substrate.
Esker	Ridge of gravel or sand laid down by a subglacial stream or one which issues from a retreating glacier.
Fan	A deposit of sand or gravel formed at the disgorgement of a river.
Fauna	The assemblage of animals of a region or period.
Flandrian	Geological stage representing the Post Glacial, Recent epoch. See Table 3.2

Flocculated	Aggregated in tufts, flakes or cloudy masses.
Fluvial	Of or belonging to rivers.
Fluvioglacial	Pertaining to glacial rivers.
Geophysical	Relating to the physics of the earth.
GLC	Greater London Council
Grab, grab-bucket	a steel bucket made in two halves hinged together, so that they dig out and enclose part of the material on which they rest; used on some dredgers and wharves.
Grit	Sandy fine gravel.
HR	Hydraulics Research Ltd.
High place value	commodities, like sand and gravel, that have a low monetary value at their pit, quarry or wharf and whose value lies in their proximity to a market ie the cost of extraction is low and therefore haulage costs will significantly affect the delivered price.
Infauna	Animals that live within a sediment.
Interstices	Small spaces between closely set sediment particles.
Isobath	A contour line of equal depth.
Jet-pump	based on the 'jet venturi' principle, water at high pressure is passed down one or two separate pipes secured to the dredge pipe. These are connected into the dredge pipe at an annular nozzle located immediately above the draghead where a vacuum is caused in the venturi tube. The jet agitates material and the loosened particles then pass up the suction pipe to the ship's hold without passing through the pump.
Jetty	a loading or unloading point formed on piling built out from river banks at right-angles to the flow.
Jurassic	Geological period, covering 195 to 135 M years ago, representing the middle of the Mesozoic era.
Lag deposit	A residual deposit consisting of the coarser, winnowed remnants of a previous sediment.
Littoral	The beach, the space between high and low tide marks, or water a little below low water mark.

Luffing boom	a conveyor on a dredger which can be raised, lowered and swivelled in a radial fashion for the unloading of a cargo.
Megaripples	Latitudinal sedimentary bed forms; small sand waves.
Metabolite	Product of metabolism.
Metalliferous	Bearing or yielding metal.
MHW	Mean High Water
MLW	Mean Low Water
Moraine	Deposit produced by the wasting (margin) of a glacier.
NBS	National Building Specification : a loose-leaf publication updated quarterly. Contains clauses of specifications from which architects or other consultants can selectively copy when writing specifications or preambles to bills of quantities.
ODN	Ordnance Datum Newlyn : sea level.
O-group	First year (of fish life).
Organomineralogic	Composed of a mixture of mineral and organic materials.
Ovigerous	Egg-carrying.
Palaeo	Concerned with the very distant past.
Periglacial	Pertaining to the regions fringing an ice sheet.
Permian	Geological period, covering 280 to 225 M years ago, representing the last part of the Palaeozoic era.
Placer	A mineral deposit formed by processes of sedimentary accumulation.
Pleistocene	Geological epoch, covering 2 M years to 10,000 years ago, representing part of the late Quaternary period (after the Pliocene). See Table 3.2
Plume	A zone of dispersion.

Proto	First formed, primitive.
Quaternary	The latest geological period, covering 2 M years ago to the present, representing the last part of the Cenozoic era.
Recent	Of the present geological epoch - Post Glacial. Also known as the Holocene. Last part of the Quaternary period. See Table 3.2
RPI	Retail Prices Index.
SAGA	Sand and Gravel Association Ltd.
Shear (stress)	Deformation in which parallel planes remain parallel, but move parallel to themselves (stress producing such deformation).
Shingle	Coarse gravel.
Side-scan sonar	A device comprising two sideways looking echo-sounders, thus using sound waves to map the acoustic properties of the sea bed surface.
Silt	Colloquially, fine residue after washing sand and gravel; specifically material in the size range 63-2µm.
SRPC	Sulphate resistant Portland Cement
SSSI	Site of Special Scientific Interest
Sub-bottom profiler	A device for detecting and mapping sound-reflecting layers in sediments below the sea bed.
Suspended load	Sediment transported within the water column.
Tectonic	Structural geology.
Tertiary	Geological period, covering 65 - 2 M years ago, representing the first (major) part of the Cenozoic era.
Till	A stiff impervious clay - boulder clay.
Warp	to turn a vessel from one place to another by hauling on warps or ropes attached to buoys, other ships, or anchor etc.
Wolstonian	Geological stage in the Pleistocene epoch, covering 200,000 to 100,000 years ago, representing the main penultimate advance of glaciers. See Table 3.2

APPENDIX 2

BIBLIOGRAPHY

ACKEFORS H. FONSELIUS S.H.
1969
Preliminary investigations on the effect of sand suction work on the bottom of Oresund.
ICES CM1969/E:13 8pp (mimeo)

ANON
1967
A sea-going dredger with an unusual pumping system.
Cement, Lime & Gravel Vol.42, No.10: 322-325

ANON
1975
Report of the Working Group on effects on fisheries of marine sand and gravel extraction.
ICES, Denmark., Co-operative Research Report No.46

ANON
1977
New dredging system for gravel. Latest developments at Reservoir Aggregates.
Quarry Management & Products July: 177-182

ANON
1979
A reply from the Sand and Gravel Dredging Industry to the I.C.E.S. Coop. Research Report No. 46.
Publication of the Sand And Gravel Assocn. of Great Britain. 32pp

ANON
1981
An aid to shipbuilding.
EEC Council Directive of 28 April 1981. (81/363/EEC)

ANON
1982
Aggregates by waterway.
Quarry Management & Products, 9 (1), 13-15.

ANON
1983
New marine aggregates depot on South Coast.
Quarry Management & Products, 10 (7), 405-410.

ANON.
1979
Reply from the sand and gravel dredging industry to the I.C.E.S. Report of the working group on effects on fisheries of marine sand and gravel extraction.
Publication of the Sand & Gravel Assoc. of Great Britain 32pp

ARCHER A.A.
1972
Sand and gravel as aggregate.
Mineral Resources Consultative Cttee. Mineral Dossier No.4, HMSO, London

ARCHER A.A.
1974
Progress and prospects of marine mining.
Mining Magazine 130 (3), 150-163.

ARTZ J.C.
1975
The economics of dredging sand and gravel for aggregate.
1st Int. Symp. on Dredging Tech., Univ.of Kent. Sept. Paper.

AUGRIS C. CRESSARD A.P.
1981
Les Granulats Marines.
CNEXO publication, Scientific & Technical Rep. No. 51.

BALL S.
1985
A high accuracy range and bearing system. Trimble Global Positioning System.
6th Mar.Meas.Forum, Haslemere.

BARNBROOK G. DORE E. et al.
1975
Concrete Practice, Cement and Concrete Assoc.

BATES A.D.
1981
Profit or loss on pre-dredging surveys.
Dredging and Port Construction. April.

BEAVER S.H.
1968
The geology of sand and gravel.
Sand & Gravel Assoc. of Great Britain publication

BELDERSON R.H. KENYON N.H. et al.
1971
Holocene sediments on the continental shelf west of the British Isles.Reprinted
from:ICSU/SCOR Working Party 31 Symposium, Cambridge 1970
The Geol.of the E.Atlantic Cont.Marg.Ed:Delaney,FM Inst.Geol.Sci.70/14:157-170

BENNELL J.D. JACKSON P.D. et al.
1982
Further developments on sea-floor geophysical probing.
Oceanology International Conference, Brighton Paper 01824.8 31pp

BLACKLEY M.W.L.
1978
Geophysical interpretation and sediment characteristics of the offshore and
foreshore areas.
Swansea Bay (Sker) Proj. Topic Rep.3 IOS Rep.No.60

BOLSTER G.C. BRIDGER J.P.
1957
Nature of spawning area of herring.
Nature, London. 179: 638pp

BOWEN D.Q.
1978
Quaternary Geology: A stratigraphic framework for multidisciplinary work.
Pergamon Press,Oxford

BOXER
1985
Land movement and coast protection.
Problems Assoc. with the Coastline; Conf. at Newport,IOW IOWCC 7pp

BRAMPTON A.H.
1985
Effects of dredging on the coast.
Problems Assoc. with the Coastline; Conf. at Newport,IOW IOWCC, 11pp

BUILDING RESEARCH ESTABLISHMENT
1982
Alkala aggregate reactions in concrete. Garston 8pp. Digest 258.

BYRD T.
1985
Road salts new suspect in AAR epidemic.
New Civil Engineer 15th August.

CARLING P.A. READER N.A.
1981
A freeze-sampling technique suitable for coarse river-bed material
Sedimentary Geology Vol.29, 233-239

CARR A.P. BLACKLEY M.W.L.
1969
Geological composition of the pebbles of Chesil Beach, Dorset.
Dorset Nat.Hist.& Arch. Soc. 90:133-40

CEMENT AND CONCRETE ASSOCIATION
1983
Minimising the risk of alkali-silica reaction: Guidance notes.
Rep.of Wkg.Pty. under Chairmanship of M.R.Hawkins.

CHAPMAN G.P. ROEDER A.R.
1969
Sea-dredged sands and gravels.
Quarry Managers' Journal 53 (7), 251-263

CHAPMAN G.P. ROEDER A.R.
1970
The effects of sea-shells in concrete aggregates.
Concrete February: 71-79.

CHESTER D.K.
1982
Predicting the quality of sand and gravel deposits in areas of fluvioglacial deposition.
Univ. of Liverpool Res. Paper No. 10.

CHILLINGWORTH P.C.H.
1979
The technology of offshore sand and gravel exploitation.
Unpublished degree thesis submitted to the Open University.

CHILLINGWORTH P.C.H.
1980
Marine sand and gravel prospects for the future.
Quarry Management & Products, 7 (4), 92-95

COLLINS M.B. HAMMOND T.M.
1979
On the threshold of transport of sand-sized sediment under the combined influence of unidirectional and oscillatory flow.
Sedimentology 26: 795-812

CONSOLIDATED GOLD FIELDS AUSTRALIA LTD. ARC MARINE LTD.
1980
Environmental impact statement. Marine Aggregate project, 3 vol.

COTTELL D.S.
1978
Some international opportunities for marine aggregates.
Trans. I.M.M. Sect.A. 87 (84-88)

COTTELL D.S.
1981
Offshore recovery of sand and gravel: An alternative to inland mining.
World Dredging & Marine Construction, November: 9-14

COUPER A.D.
1985
The marine boundaries of the United Kingdom and the law of the sea
Geogrl.J. 151 (2), 228-236

CRUICKSHANK M.J.
1964
Mining offshore alluvials
Symp.on opencast mining quarrying & alluvial mining Paper 7: 125-155

CRUICKSHANK M.J.
1982
Recent studies on marine mineral resources
Oceanology Internat.Conf.papers Vol.1,Paper No.01821.2 18pp

CRUICKSHANK M.J. ROMANOWITZ C.M. et al.
1968
Offshore mining present and future.
Engineering & Mining J. January 84-91

CULLINGFORD R.A. SMITH D.E.
1966
Late glacial shorelines in eastern Fife.
Transactions of the Inst. of Brit. Geographers 39: 31-51

DE GROOT S.J.
1979
An assessment of the potential environmental impact of large-scale sand-dredging for the building of artificial islands in the North Sea.
Ocean Management 5, 211-232

DE GROOT S.J.
1979
The potential environmental impact of marine gravel extraction in the N. Sea
Ocean Management 5: 233-249

DE GROOT S.J.
1981
Bibliography of literature dealing with the effects of marine sand and gravel extraction on fisheries.
ICES Rep. CM1981/E:5 39pp

DEEGAN C.E. KIRBY R. et al.
1973
The superficial deposits of the Firth of Clyde and its sea lochs.
Rep. Inst. Geol. Sci. Vol.73/9 1-44.

DEPARTMENT OF THE ENVIRONMENT
1976
Aggregates: the way ahead.
Report of the Verney Committee.

DEPARTMENT OF THE ENVIRONMENT
1978
Report of the Advisory Committee on Aggregates,
DoE Circular 50/78, Welsh Office Circular 92/78

DEPARTMENT OF THE ENVIRONMENT
1982
Guidelines for Aggregate Provision in England and Wales.
DoE Circular 21/82, Welsh Office Circular 30/82

DEPARTMENT OF THE ENVIRONMENT
1986
A Description of the United Kingdom Marine Dredged Aggregates Industry: 1985
NUNNY R.S. & CHILLINGWORTH P.C.H. Min. Plan. Res. Proj. Rept. PECD 7/1/163

DEPARTMENT OF TRANSPORT
1976
Specification for road and bridge works.

DEPARTMENT OF TRANSPORT
1978
Specification for road and bridge works.
Supplement No. 1.

DEPARTMENT OF TRANSPORT
1982
Freight facilities grants; December.
(also Scottish Development Dept. and Welsh Office).

DICKSON R.R.
1975
A review of current European research into the effects of offshore mining on the fisheries.
Offshore Technology Conf. Paper No.OTC 2159

DICKSON R.R. LEE A.
1972
Study of effects of marine gravel extraction on the topography of the sea bed.
I.C.E.S. CM1972/E:25

DICKSON R.R. LEE A.
1973
Gravel extraction: effects on seabed topography.
Offshore Services Vol.6,No.6 32-39 & Vol.6,No.7 56-61

DINGLE R.V.
1970
Quaternary sediments and erosional features off the north Yorkshire coast, western North Sea.
Marine Geology 9(1970) M17-M22

DOBSON M.R. EVANS W.E. et al.
1971
The sediment on the floor of the southern Irish Sea.
Sea. Marine Geol. 11: 27-69

DONOVAN D.T.
1973
The geology and origin of the Silver Pit and other closed basins in the North Sea.
Proc. York. Geol. Soc. Vol.39, Pt.2, No.13 267-293

DRAKE D.E.
1976
Suspended sediment transport and mud deposition on continental shelves.
Chapter 9 IN
Mar.Sed.Trans. & Environ. Mngmt. Eds:Stanley & Swift. John Wiley & Sons,127-155

DRAPER L.
1967
Wave activity at the sea-bed around north western Europe.
Marine Geology Vol.5: 133-140

DYER K.R.
1985
Sediment transport in the Solent and around the Isle of Wight.
Problems assoc. with the coastline; Conf.at Newport,IOW IOWCC 6pp

D'OLIER B.
1975
Some aspects of late Pleistocene-Holocene drainage of the River Thames in the eastern part of the London Basin.
Phil.Trans. R. Soc. Lond. A.279, 269-277

EAGLE R.A. HARDIMAN P.A. et al.
1978
The field assessment of effects of dumping wastes at sea: 3 A survey of the sewage sludge disposal area in Lyme Bay.
Fish. Res. Tech. Rep. No.49, MAFF Dir. Fish. Res., Lowestoft

EAGLE R.A. HARDIMAN P.A. et al.
1979
The field assessment of effects of dumping wastes at sea: 4 A survey of the sewage sludge disposal area off Plymouth
Fish.Res.Tech.Rep. No.50, MAFF Dir.Fish.Res., Lowestoft

EDEN R.A.
1970
Marine gravel prospects in Scottish waters.
Cement, Lime & Gravel September, 237-240

EDWARDS A.G.
1963
Recent work on aggregates of the Scottish building research station
Sand & Gravel Assoc.: Fourth short course, Imperial College, London.

EDWARDS A.G.
1970
Scottish aggregates: their suitability for concrete with regard to rock constituents.
Building Research Station Paper C.P. 29/70 (Watford)

EISMA D.
1975
Holocene sedimentation in the Outer Silver Pit area (southern North Sea).
Marine Science Communications 1(6), 407-426

EVANS C.D.R.
1982
The geology and superficial sediments of the inner Bristol Channel and Severn Estuary.
Severn Barrage, proc. of symposium org. by I.C.E. 35-49

EVANS D.
1984
Catalogue of BGS offshore maps with selected reports and publications.
British Geological Survey,Mar.Geol.Res.Prog.Int.Rep. 84/7 25pp.

FLEMING N. STRIDE A.
1967
Basal sand and gravel patches with separate indications of tidal current and storm wave paths, near Plymouth.
J. Mar. Biol. Assoc. of U.K. 47:433-444

GARRARD R.A.
1977
The sediments of the South Irish Sea and Nymphe Bank area of the Celtic Sea IN Quat.Hist of the Irish Sea, Eds:Kidson,C & Tooley,M.J. Seel Hse Press, Livpl.

GEOLOGICAL SURVEY OF JAPAN
1975
Exploitation of offshore sand and gravel in Japan.
Proc.20th session, c.c.o.p., Tokyo.Doc. c.c.o.p. (XII)/24, 288-293.

GEORGE C.L. WARWICK R.M.
1985
Annual macrofauna production in a hard-bottom reef community.
J. Marine Biological Association of U.K. 65

GLASBY G.P.
1982
Marine mining and mineral research activities in Europe.
Marine Mining Vol.3, Nos.3-4: 379-409

GRANT C.L. et al.
1974
A study to understand the environmental and ecological impact of marine sand and gravel mining order to prepare guide lines for mining operations in the sea.
National Technical Information Service, COM-75-10714:83pp

GRAVESTOCK M.J.
1985
Administration of offshore dredging.
Problems Assoc. with the Coastline; Conf. at Newport, IOW. IOWCC

GRAY J.S.
1974
Animal-sediment relationships
Oceanogr. Mar. Biol: Ann. Rev. 12: 223-261

GREATER LONDON COUNCIL
1984
Specification for concrete.
Architect's Department, GLC.

HALLERMEIER R.J.
1981
Seaward limit of significant sand transport by waves: An annual zonation for seasonal profiles.
Coastal Eng. Tech. Aid No.81-2. US Army, Corps. of Engineers. 18pp.

HALLIDAY J.S. MULLETT J.C.
1983
Future prospects for the commercial exploitation of marine sands from Liverpool Bay.
Univ.of Liverpool, Dept. of Industrial Studies.

HAMMOND S.D.C. HEATHERSHAW A.D. et al.
1984
A comparison between Shield's threshold ctirerion and the movement of loosely packed gravel in a tidal channel.
Sedimentology 31; 51-62.

HARRIS L.G.
1976
Environmental assessment of sand and gravel marine mining.
Proc.of the mar.minerals wkshp.,Silver Spring,Md., NOAA S/T 76-2484 73-79

HASSELMANN K. ET AL
1973
Measurements of wind-wave growth and swell decay during the Joint North Sea Wave Project (JONSWAP).
Deutsche Hydrographische Zeitschrift Supplement A, 8, No. 12

HEATHERSHAW A.D. CARR A.P. et al.
1981
Final Report: Coastal erosion and nearshore sedimentation processes. Swansea Bay (Sker) Project.
Topic Report B I.O.S. Report No.118 67pp

HERBICH J.B.
1976
Technology of sand, gravel and shell marine mining.
Proc.of the mar.minerals wkshp.,Silver Spring,Md., NOAA S/T 76-2484

HESS H.D.
1971
Marine sand and gravel mining industry of the U.K.
Nat.Ocean.& Atmos.Admin.Tech.Rep. ERL 213-MMTC 1,U.S.Dept.of Commerce,Colorado.

HILLING D.
1981
Waterway transport of industrial minerals - a neglected mode.
'IM' Mineral Transportation Survey '81, March. 17-22.

HOBBS D.W.
1982
Mortar prisms - reactive aggregate.
Mag. of Conc. Res. 34 (119).

HORIKAWA K. SASAKI T. et al.
1977
Mathematical and laboratory models of shoreline changes due to dredged holes.
J. of Faculty of Eng., Univ. of Tokyo (B) Vol.XXXIV, No.1, 49-57

HOUBOLT J.J.H.
1968
Geology of the continental shelf around Britain. A survey of progress IN
Geology & Shelf Seas Ed:Donovan,D.T., Oliver & Boyd, Edinburgh & London

HOWARD A.E.
1982
The distribution and behaviour of ovigerous edible crabs (cancer pagurus) and consequent sampling bias.
J. Cons. int. Explor. Mer. 40: 259-261

HYDRAULICS RESEARCH
1972
The movement of offshore shingle, Worthing.
H.R. Wallingford, Rep. EX591

HYDRAULICS RESEARCH
1976
The effect in coastline changes of wave refraction over dredged areas.
H.R. Wallingford, Rep. EX728

HYDRAULICS RESEARCH
1977
Solent Bank, Pot Bank & Prince Consort dredging.
H.R. Wallingford, Rep. EX770

HYDRAULICS RESEARCH
1983
Wave attenuation over uneven seabed topography: A study of wave changes in relation to offshore dredging.
H.R. Wallingford, Rep. EX1143

HYDRAULICS RESEARCH
1984
South East Isle of Wight: Effects of proposed dredging.
H.R. Wallingford, Rep. EX1208

HYDRAULICS RESEARCH
1984
Sand and Gravel mobility in relation to offshore dredging. A radioactive tracer study in the Shipway Channel off the Suffolk coast.
H.R. Wallingford, Rep. EX1176

I.C.E.
1985
Research requirements in siltation, dredging and dispersion. Appendix B.
Res.Requirements in Coastal Eng. Rep.by Coast.Eng.Res.Panel & Work.Parties 27-28

I.C.E.S.
1975
Report of the Working Group on effects on fisheries of marine sand and gravel extraction.
Co-operative Res. Rep. 46

I.C.E.S.
1977
Second report of the ICES Working Group on effects on fisheries of marine sand and gravel extraction.
Co-operative Res. Rep. 64

I.C.E.S.
1979
3rd mtg. of the Working Group. Report of the I.C.E.S. Working Group on effects on fisheries of marine sand and gravel extraction.
CM1979/E:3

JAFFREY L.J. MOTYKA J.M. et al.
1978
The effect of offshore dredging on coastlines.
Proc. of 16th Coast. Eng. Conf. Vol.2: 1347-1358

JARDINE F.
1985
Alkali-silica: Reactions or over-reactions?
SAGA Bulletin 17 (3), 3-4.

JORDAN J.P.R. PIKE D.C.
1976
Proposals in relation to limits on chloride content of marine aggregates.
SAGA Tech. Paper No.11,

KAPLAN E.H. WELKER J.R. et al.
1974
Some effects of dredging on populations of macrobenthic organisms.
Fish.Bull., Fish. Wildlife Serv. U.S. 72: 445-480

KELLAND N.C.
1971
Exploration techniques for gravel.
J. Geol. Soc. Lond. Vol.127, page 531

KENYON N.H.
1970
Sand ribbons of European tidal seas.
Marine Geol. 9: 25-39

KENYON N.H. STRIDE A.H.
1970
The tide-swept continental shelf sediments between the Shetland Isles & France.
Sedimentology 14:159-173

KIDSON C. CARR A.P.
1959
The movement of shingle over the seabed close inshore.
Geol.Journal 125:380-389

KING C.A.M.
1972
Beaches and Coasts. Edward Arnold, London. 570pp.

KIRBY R. PARKER W.R.
1983
Distribution and behaviour of fine sediment in the Severn Estuary and inner Bristol Channel, U.K.
Canadian J. of Fish & Aquat.Sci. Vol.40 (Supp.1)

KOMAR P.D.
1976
Boundary layer flow under steady unidirectional currents. Chapter 7 IN
Mar.Sed.Trans.& Environ.Mngmt.Eds:Stanley & Swift. John Wiley & Sons. 91-106

LARSONNEUR C.
1965
Recherches sedimentologiques et geologiques en Manche Centrale.
Rev.Trav.Inst.Peches Maritimes 29:225-242

LEES B.J.
1983
Final Report: A study of nearshore sediment transport processes. Sizewell-Dunwich Banks field study.
Topic Report 7 I.O.S. Report No.146 43pp.

LUTTIG G.
1978
Present and future economic value of marine sand and gravel.
Industrial Minerals No.133: 53-57

MAFF
1981
Code of practice for the extraction of marine aggregates.
Jointly prepared by SAGA, CEC, MAFF & DAFS, December.

MAFF
1985
Ministerial Information in MAFF (Minim)

MARTIN D.(ed)
1984
Specification 85. Building methods and products.
2 Technical and Product Information. Architectural Press.

MAY E.B.
1973
Environmental effects of hydraulic dredging in estuaries.
Ala.Mar.Resour.Bull. 10: 1-8

MCCAVE I.N.
1971
Sand waves in the North Sea off the coast of Holland.
Marine Geology Vol.10, 199-225

MCCAVE I.N. CASTON V.N.D. et al.
1977
The Quaternary of the North Sea IN
British Quaternary Studies - Recent Advances. Ed: Shotton, F.W. O.U.P. 187-204

MCLELLAN A.C.
1967
The distribution of sand and gravel deposits in West Central Scotland and some problems concerning their utilisation.
Univ. of Glasgow, Dept.of Geography.

MILES A.J.
1985
The marine sand and gravel industry of the U.K.
World Dredging & Marine Construction August: 29-32

MILLER J.M. ROBERTS P.D. et al.
1977
Towed seabed gamma ray spectrometer for continental shelf surveys.
IAEA, Int.Symp.on Nuc.Tech.in Explor.,Extract.& Proc.of Min.Reserves SM216162

MILLNER R.S. DICKSON R.R. et al.
1977
Physical and biological studies of a dredging ground off the East Coast of England.
I.C.E.S. CM1977/E:48

MIN.OF HOUSING AND LOCAL GOVERNMENT
1948
Report of the Advisory Committee on sand and gravel.
(The Waters Report)

MOORE P.G.
1977
Inorganic particulate suspensions in the sea and their effects on marine animals.
Oceanogr.Mar.Biol: Ann.Rev. 15:225-363

MOTTERSHEAD D.N.
1977
The Quaternary evolution of the south coast of England.
Quat.Hist. of the Irish Sea, Eds: Kidson,C & Tooley,M.J. Seel Hse Press, Livpl.

MOTYKA J.M. WILLIS D.H.
1974
The effect of wave refraction over dredged holes.
Proc. of 14th Coast. Eng. Conf. Copenhagen, Vol.1

MURRAY L.A. NORTON M.G.
1979
The composition of dredged spoils dumped at sea from England and Wales.
Fish.Res.Tech.Rep. No.52, MAFF Dir.Fish.Res., Lowestoft

MURRAY L.A. NORTON M.G. et al.
1980
The field assessment of effects of dumping wastes at sea: 7 sewage sludge and industrial waste disposal in the Bristol Channel.
Fish. Res. Tech. Rep. No.59, MAFF Dir.Fish.Res., Lowestoft

NATURE CONSERVANCY COUNCIL
1979
Nature conservation in the marine environment.
Report of the NCC/NERC Joint Working Party

NEBRIJA E.L. WELKIE C.J. et al.
1978
Geophysical-geological exploration and evaluation of offshore sand and gravel deposits.
10th Annual Offshore Technology Conference, Houston, Texas. 1187-1198.

NIXON P.J. BOLLINGHAUS R.
1983
Testing for alkali reactive aggregates in the UK.
Proc.6th Int.Conf.on Alkalis in Conc:Res.&Prac/Tech.Univ.Denmark/Copenhagen/June

NOAKES J.E. HARDING J.L.
1982
Nuclear techniques for sea floor mineral exploration.
Oceanology Internat.Conf.papers Vol.1,Paper No.01821.3 16pp

NORTON M.G. FRANKLIN A. et al.
1984
The field assessment of effects of dumping wastes at sea:12 The disposal of sewage sludge, industrial wastes and dredged spoils in Liverpool Bay.
Fish.Res.Tech.Rep. No.76,MAFF Dir.Fish.Res., Lowestoft. 50pp

OELE E.
1971
The Quaternary geology of the southern area of the Dutch part of the North Sea.
Geol.Mijnb. 50: 461-575

O'CONNOR B.A.
1985
Coastal sediment modelling.
Offshore Coast.Modlng,Eds:Dyke,Moscardin & Robson, Spinnger-Verlag,Berlin/N.Y.

PANTIN H.M.
1977
Quaternary sediments of the northern Irish Sea IN
Quat.Hist of the Irish Sea, Eds:Kidson,C & Tooley,M.J. Seel Hse Press, Livpl.

PANTIN H.M.
1978
Quaternary sediments from the north-east Irish Sea:Isle of Man to Cumbria.
Bull. Geol. Surv. G.B. No.64

PASHO D.W.
1986
The UK offshore aggregate industry: A review of management practises and issues.
Canada Oil & Gas Lands Admin., Ocean Mining Div.

PEARSON E.A. STORRS P.N. et al.
1967
Some physical parameters and their significance in marine waste disposal IN
Pollution and Marine Ecology Eds:Olson & Burgess 297-316, Interscience N.Y.

POLLOCK D.T.
1985
Marine dredged aggregates.
Mineral Planning, 24: 19-20.

RHOADS D.C.
1974
Organism-sediment relations on the muddy sea floor.
Oceanogr. Mar. Biol: Ann.Rev. 12:263-300

RILEY J.D. SYMONDS D.J. et al.
1981
On the factors influencing the distribution of O-group demersal fish in coastal waters.
Rapp. P.-v. Reun. Cons. int. Explor. Mer 178: 223-228

ROWE R.P.
1976
Dredging for marine aggregates in the U.K.
Harbour & Port Construction February: 15-19

ROYAL COMM.ON COASTAL EROSION
1907-1911
 Parts 1,2 & 3.

SAUNT T.
1974
The distribution of quarry products - transport of aggregates by sea.
Quarry Managers' Journal March: 83-92

SCOTTISH DEVELOPMENT DEPT.
1977
National Planning Guidelines: Aggregate Working.

SCOTTISH DEVELOPMENT DEPT.
1981
National Planning Guidelines: Priorities for development planning.

SEYMOUR R.J.
1977
Estimating wave generation on restricted fetches.
Jul. ASCE Vol.103, No.WW2

SHELTON R.G.J. ROLFE M.S.
1972
The biological implications of aggregate extraction: recent studies in the English Channel.
I.C.E.S. CM1972/E:26

SIBTHORP M.(ed)
1975
The North Sea - challange and opportunity.
Rep.of Study Group of D.Davis Memorial Inst.of Int.Studies/Europa Publcs/London.

SIMS L.
1983
The influence of ground granulated blast furnace slag on the alkali-reactivity of flint aggregate concrete in the U.K.
Proc.6th Int.Conf.on Alkalis in Conc:Res & Pract./Tech.Univ.Denmark/Copenhagen

SINGLETON D.F.
1981
Transportation of bulk materials to the construction industry.
'IM' Mineral Transportation Survey '81, March: 33-35.

SNOWDON L.C. EDWARDS A.G.
1962
The moisture movement of natural aggregate and its effect on concrete.
Mag. Concr. Res. 14 (49), 109-115

SPREULL W.J. UREN J.M.L.
1986
Marine aggregates and aspects of their use especially in South East of England.
Joint publication of St'. Albans Sand & Gravel Co.Ltd. & Civil & Marine Ltd.

STEERS J.A.
1964
The Coastline of England and Wales (2nd Ed.) C.U.P., Cambridge 780pp

STRIDE A.H. BELDERSON R.H. et al.
1972
Longitudinal furrows and depositional sand bodies of the English Channel.
Extrait due Memoire du B.R.G.M. No.79

TALBOT J.W. TALBOT G.A.
1974
Diffusion in shallow seas and in English coastal and estuarine waters.
Rapp. P.-v. Reun. Cons. int Explor. Mer. 167: 93-110

THOMPSON M.E.
1973
Sand plus gravel deposits in the central part of the Firth of Forth.
BGS Marine Geology Internal Report No. 73/5

TINSEY D.
1983
The trade in sea-dredged aggregates to the Thames.
Port of London. First Edition: 29-33

TROY J.F.
1985
Concrete and aggregates testing standards in 1985. Paper presented at mtg.of London & Home Counties Branch of the Inst. of Quarrying, 7th March.

VEENSTRA H.J.
1969
Gravels of the southern North Sea
Sea. Mar. Geol. 7, 443-64

VERBECK G.J.
1975
Mechanisms of corrosion of steel in concrete.
ACI Special Publication SP-49, 21-38.

WALKOTTEN W.J.
1976
An improved technique for freeze sampling stream-bed sediments.
Pacific N.W. Forest & Range Exper. Station, USDA Forest Service Research Note

WEBB P.
1979
Legal and economic aspects of dredging marine aggregates in the U.K.
World Dredging & Marine Construction November: 34-38

WENTWORTH C.K.
1922
A scale of grade and class terms for clastic sediments.
J. Geol. 30: 377-392.

WHEELER B.A.
1980
ARC Marine expanding in lucrative U.K. offshore aggregate market.
World Dredging & Marine Construction November: 21-24

WHITE I.C. ROLFE M.S. et al.
1974
Disposal of wastes at sea, Part III. The field assessment of effects.
ICES CM1974/E:25 14pp (mimeo)

WILKES B.ST.J
1971
Nautical Archaeology David & Charles. 294pp.

WILLIAMS S.J. KIRBY R. et al.
1982
Sedimentation studies relevant to low level radioactive effluent dispersal in the Irish Sea, Pt.II
Seabed morph, seds. & shallow sub-bottom stratig. of E.Irish Sea IOS Rep.

WRIGHT P. CROSS J.S. et al.
1978
Shingle tracing by a new technique.
Proc. of the 16th International Conf. on Coastal Engineering 10pp

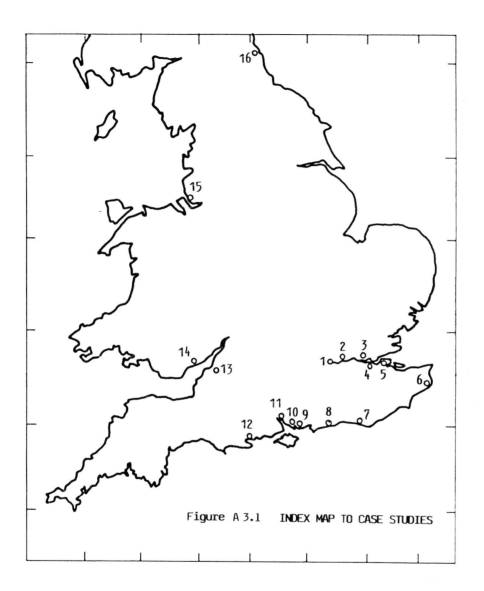

Figure A 3.1 INDEX MAP TO CASE STUDIES

APPENDIX 3
WHARF CASE STUDIES

Eighteen wharves around the UK illustrate a range of issues associated with their location and operation. These include wharves that are: small and conjested in urban areas to large wharves in rural areas; old and new ones; some that are beset with navigation problems; and those that have rail links.

One case study (No 12) is based on an Inspector's report of a public inquiry where the developer decided not to implement the permission.

Each case study follows a standard format of presentation to enable quick comparisons to be made between wharves. Unfortunately three wharves have only limited information.

All location maps in this Appendix are on a scale of 1:10,000.

WHARF: Cringle, Nine Elms Lane, Wandsworth, London.   **WHARF AREA** 0.81 ha.   **NGR** TQ 293776

OPERATOR: Metro-Greenham Aggregates Ltd. (a jointly owned company by RMC and Taylor Woodrow).

LAND OWNER: Metro-Greenham Aggregates Ltd.

SITE DESCRIPTION: Located a short distance downstream of Battersea Power Station in Chelsea Reach on the right (south) bank of the River Thames. To the west of the site is a new GLC waste transfer station (road to barge). Immediately to the east is a DoE museum storage building with Nine Elms Industrial Estate beyond. The area generally is in a zone of re-development with some parts demolished in readiness for new development. Opposite the site, on the other side of the river are residential flats.

COMPANY DELIVERING MARINE DREDGED MATERIAL: East Coast Aggragates (RMC), formerly British Dredging Co.

MATERIALS LANDED: Sand and gravel from southern North Sea.

DATE LANDINGS OF MARINE DREDGED MATERIAL COMMENCED: 1964

OWNERS/OPERATORS POLICY TOWARDS WHARF: To maximise use of wharf for marine aggregate processing and distribution.

OPERATIONAL CONSIDERATIONS:

Berth: Jetty 61m long set about 25m into river. Three specially designed vessels berth ('flatties' - with lowering masts and low profile superstructure to negotiate Thames bridges). These are the Bowsprite, Bowbelle and Bowtrader whose cargo carrying capacities range from 1,700 - 2,000 tonnes. Vessels lie over tide and sail on following high tide.

Wharf Case Study No 1

Cringle Wharf, Battersea, London

This site is not only the furthest upstream on the River Thames receiving marine sand and gravel but is also the only wharf for that purpose above Tower Bridge. Navigation beneath the nine bridges after Tower Bridge involves specially designed vessels known as 'flatties'.

Discharging from mv Bowtrader

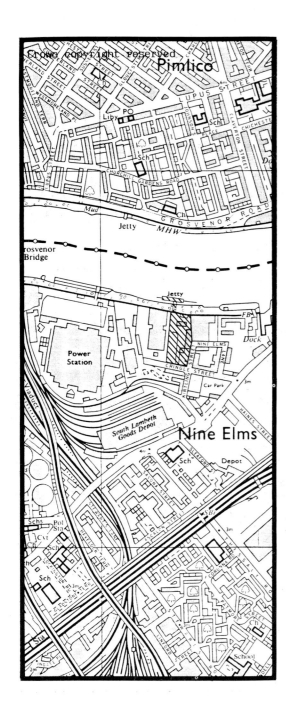

Discharge:	Two rail mounted grab cranes (one used for standby purposes only) capable of lifting 5½ tonnes in each load. Discharge takes between 5 to 7 hours depending on size of cargo to be unloaded.
Wharfage/Plant:	Washing, grading and crushing plant rated at 350t/hr. Maximum annual capacity is around 650,000 tonnes but the current output is under 500,000 tonnes.

Products : 40mm, 20mm, 10mm and 5mm medium sand.

Processed material sold to Greenham Concrete Ltd who have a batching plant on site. Remaining washed aggregate sold to other concrete companies, contractors and builder's merchants etc.

Source of wash water : River Thames under Water Authority licence. Chloride content satisfactory. Regular monitoring ensures that quality is maintained.
Flocculent used to settle silt. Some silt is removed for off-site disposal at a tip and the remainder is sold as reject sand used for fill.

Stockpiles : 16,000 tonnes (unprocessed) - separate stocking bays for BAD and dredged sand.
2,000 tonnes (processed).

Material Quality:	No shell or chloride problems. Some user prejudice experienced in the past but is now less common once customers become more informed. About 70% of material goes into concrete batching plants, the remainder is sold as washed aggregate with a little sold as fill (BAD).
Market penetration:	Normally 10 radial miles on both sides of the Thames, although there is less congestion in South London which makes marketing easier.

PLANNING CONSIDERATIONS:

    No restrictions on hours of working.
    Good access to main road network.
    Site located in Nine Elms Industrial Area.

COMMENT:

    The Cringle Wharf is the sole example of marine sand and gravel landings above the Thames bridges. It is well placed to serve an important market.

Plant site from Cringle Street

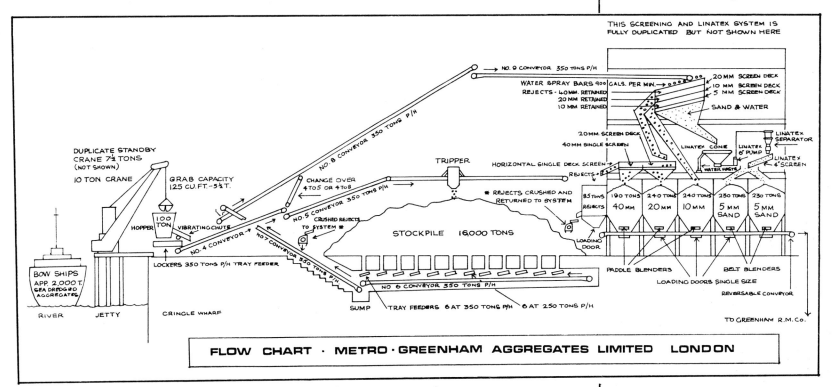

FLOW CHART · METRO · GREENHAM AGGREGATES LIMITED LONDON

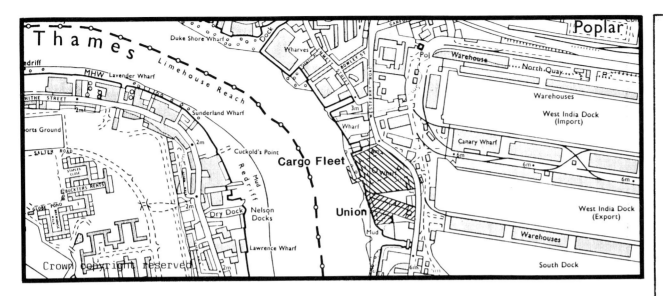

Wharf Case Studies No 2 A & B

Cargo Fleet and Union Wharves
Westferry Road, Millwall, London

These two wharves, which adjoin each other, are located in Limehouse Reach of the River Thames. The area is in the heart of the dockland which is undergoing major re-development.

Union Wharf (right) and part of Cargo Fleet (left)

Cargo Fleet

Cargo Fleet

Union

CASE STUDY No 2A

WHARF: Cargo Fleet, Westferry Road, Tower Hamlets, London.   **WHARF AREA** 1.48 ha   **NGR** TQ 370803

OPERATOR: Tarmac Roadstone Ltd (Southern)

LAND OWNER: Tarmac PLC

SITE DESCRIPTION: Wharf located on the left (north) bank of the River Thames in dockland area. Another marine sand and gravel operation adjoins to the south and a yacht repair yard adjoins to the north.

COMPANY DELIVERING MARINE DREDGED MATERIAL: Tarmac (Marine) Ltd.

MATERIALS LANDED: Sand and gravel from southern North Sea (Shipwash Bank).

DATE LANDINGS OF MARINE DREDGED MATERIAL COMMENCED: Site acquired by former Hoveringham Group in 1966. Landings commenced around 1968/69.

OWNERS/OPERATORS POLICY TOWARDS WHARF: To maximise area for the reception and processing of marine sand and gravel.

OPERATIONAL CONSIDERATIONS:

Berth: Jetty : 50m long set about 20 metres into river.

Vessel normally arrives 3 hours before high water and leaves as the tide turns. Vessel would ground if it stayed over low water. Berth periodically dredged by private contractor largely because of river silt.

mv Hoveringham VI

Self discharge to wharf

Overhead tripper and raw stockpile

Vehicle loading

Discharge: Self discharge with drag scraper @ 1,000t/hr from Hoveringham VI. Discharge completed in 2 ½ hours.

Wharfage/Plant: Washing, grading and crushing plant.
Processing plant rated at 200 t/hr with a capability of producing up to 500,000 tonnes per annum.
Concrete batching plant (Tarmac Topmix).

Products : 40mm, 20mm, 10mm, 5mm (medium B.S. sand)

Stockpiles : 15,000t unprocessed
45,000t processed

Source of washing water : River Thames.

Silt periodically removed from site by tanker for tip disposal.

Material Quality: No shell or chloride problems : washed material complies with British Standard. Quality of dredged aggregate good. No user prejudice noticed.

Market penetration: Both sides of River Thames, up to 20 radial km.

PLANNING CONSIDERATIONS:

None.

Access to A13 reasonable (improvements to Westferry Road are proposed by the London Dockland Development Corporation with re-alignment taking part of wharf area. Road scheme will not affect operating area).

COMMENT: The wharf is well placed to serve London and especially the docklands re-development area.

CASE STUDY No 2B

WHARF: Union, Westferry Road, Tower Hamlets, London. **WHARF AREA** 0.78 ha **NGR** TQ 370802

OPERATOR: Pioneer Aggregates (Millwall) Limited.
A wholly owned subsidiary of Pioneer Aggregates (UK) Ltd.

LAND OWNER: British Dredging Services Ltd.

SITE DESCRIPTION: Wharf located on the left (east) bank of the River Thames in the dockland area. Another marine sand and gravel operation adjoins to the north.

COMPANY DELIVERING MARINE DREDGED MATERIAL: British Dredging Aggregates Limited

MATERIALS LANDED: Sand and gravel from southern North Sea.

DATE LANDINGS OF MARINE DREDGED MATERIAL COMMENCED: 1967 (as British Dredging)
1976 (as BD - Pioneer)
1985 (as Pioneer)

OWNERS/OPERATORS POLICY TOWARDS WHARF: To make full use of the space available for processing/storing marine dredged material.

OPERATIONAL CONSIDERATIONS:

Berth: Length of berth: 95m

Berthing of vessels normally takes place between 2 hours before high water and one hour after high water.

Maximum size vessel : 3,750 tonnes (Bowknight). Berth dredged by a private contractor when river silt builds up. N.B. This is noticeable when dredgers start to slip on their berth.

Plant area

Raw feed to plant

Loading processed gravel

Silt settlement plant

Plant area

Discharge: Self discharge facilities. Discharge from Bowknight takes 3 hours and from Bowstream (1,800t) it takes 2 hours.

Wharfage/Plant: Washing, grading and crushing plant.
Plant rated at about 170 t/hr equivalent to around 400,000 t.p.a.
Concrete batching plant supplied on adjoining site.

Products : 40mm, 20mm, 10mm, 5mm (medium B.S. sand)

Stockpiles : 15,000t unprocessed
5,000t processed

Source of wash water : mains.

A Mecatex silt dewatering plant is incorporated into the washing system which flocculates the silt particles to separate water.

Silt removed for off-site disposal at a tip.

Material Quality: No chloride or shell problems. The majority of the material goes to supply concrete batching plants.

Market penetration: 30 radial km. Penetration south of the Thames is via London Bridge or the Blackwall Tunnel.

PLANNING CONSIDERATIONS:

No restrictions on hours of working.

Access to A13 reasonable (improvements to Westferry Road are proposed by the London Dockland Development Corporation with re-alignment near the wharf).

COMMENT: The wharf is well placed to serve London and especially the docklands re-development area.

WHARF: Purfleet, Thurrock, Essex   **WHARF AREA:** 2.5 ha   **NGR:** TQ 576767

OPERATOR: Purfleet Aggregates Ltd [a joint company formed by Civil & Marine Ltd and RMC (UK) plc] to process and market marine sand and gravel.

LAND OWNER: Civil & Marine Limited

SITE DISCRIPTION: A new wharf located in Long Reach on the left (north) bank of River Thames in an industrial area, immediately west of the Dartford Tunnel northern portal. The site has a processing plant for dredged material and value-added plant.

COMPANY DELIVERING MARINE DREDGED MATERIAL: Civil & Marine Ltd

MATERIALS LANDED: Sand and gravel mainly from Cross Sands off Lowestoft; slag (to meet type 1 road specification) from Holland; and granulated slag (for cement replacement) from France and Belgium.

DATE LANDINGS OF MARINE DREDGED MATERIAL COMMENCED: June 1981

OWNERS/OPERATORS POLICY TOWARDS WHARF: To maximise sales of landed materials by leasing parts of site to other companies using these materials.

OPERATIONAL CONSIDERATIONS:

Berth:
Vessel size limited to 5,200 dwt
Size of vessel normally used : 4,500 dwt
Depth of berth water : 7.0m at MLWS
Length of mooring jetty : 45m

Discharge:
Discharge of cargoes can take place at all states of tide. Self-discharge from vessel luffing boom via jetty hopper having an adjustable level depending on state of tide. Material

---

Wharf Case Study No 3

Purfleet Wharf, Purfleet, Essex

This is a new wharf strategically located alongside the M25 with potential for a rail link in the future. Various value added plant on site.

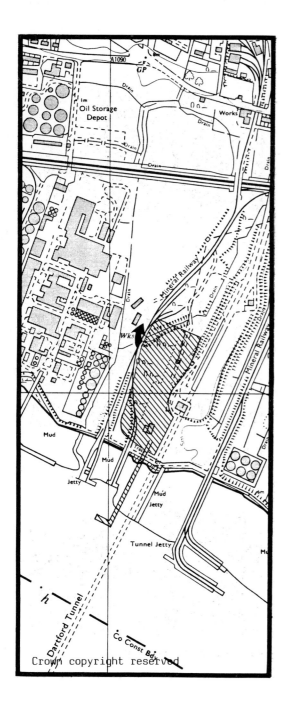

conveyed along 280m jetty to plant area. Landing and processing of materials can take place 24 hours per day, 7 days per week depending on demand.

Wharfage/Plant: Sand and gravel processed by washing, grading and crushing through a plant rated at 300 tph; 400,000 tpa; (operational from August 1984).

Products : 40mm, 20mm, 10mm and 5mm (medium sand)

Processed material either sold ex-wharf or used in concrete manufacture:batching plant on site (operational from December 1984). Material also used in asphalt manufacture:plant on site.

Source of wash water : mains, make up recirculating.

Means of discharge of silt laden water : settlement in holding tank. Closed-circuit operation. Solids periodically removed for off-site disposal.

Stockpiles : 30,000 tonnes (unprocessed)
12,000 tonnes (processed). Equivalent to ten days supply.

Also on site Civil & Marine Ltd have a grinding mill to process and market granulated slag. Tilbury Roadstone Ltd purchase marine sand and gravel from Purfleet Aggregates Ltd for use in the manufacture of asphalt.

Material Quality: Both chloride and shell content satisfy limitations in BS 882:1983.

Market penetration: Currently up to 25 km with competition from land-won sources. The opening of the M25 northbound now offers potential for expansion in the future if land-won supplies are insufficient to meet demand. However, southbound on the M25, the Dartford Tunnel is seen as an inhibiting factor due to long traffic delays coupled with other landings of marine sand and gravel at Erith, Greenhithe and Dartford.

PLANNING CONSIDERATIONS:

Numerous planning permisssions have been granted by Thurrock Borough Council for the wharf and ancillary development.

Site adjoins M25 by Dartford Tunnel with junction 31, within two kms. Haul route to motorway and A12 through an industrial area. No residential development.

Existing private railway line adjoins site. Possible rail link should demand arise. Space reserved on wharf for sidings. Owner of line and British Rail are aware of potential use.

COMMENT:

The development of this new wharf by Civil and Marine Ltd in conjunction with RMC Group Plc was taken in the expectation that supplies of land-won sand and gravel would diminish. The short-term stategy is to serve local Essex markets extending in the medium term to greater movement along the primary road network and in the long-term by a possible rail link into the site from adjoining loop line on wharf boundary.

Processing plant

View looking northwards

Concrete batching plant

WHARF: Northfleet, Botany Marshes, Kent. **WHARF AREA** 5.2 ha **NGR** TQ 611759

OPERATOR: Hall Aggregates (South East) Ltd.

LAND OWNER: Hall Aggregates (South East) Ltd.

SITE DESCRIPTION: Located in Northfleet Hope on the right (south) bank of the River Thames on Botany Marshes.

COMPANY DELIVERING MARINE DREDGED MATERIAL: South Coast Shipping (RMC).

MATERIALS LANDED: Sand and gravel from southern North Sea.

DATE LANDINGS OF MARINE DREDGED MATERIAL COMMENCED: 1967.

OWNERS/OPERATORS POLICY TOWARDS WHARF: To optimise use of wharf for sea dredged material.

OPERATIONAL CONSIDERATIONS:

Berth: Vessels berth alongside jetty dolphins 61 metres long set 274 metres from the shore.

Discharge: Pump from ship's hold using river water to slurry cargo. A sand cargo of 4,000 tonnes will take one hour to discharge whereas a mixed cargo will take 3 to 4 hours to discharge (the stonier the cargo the longer the discharge time). Flexible coupling slots onto ship's deck to link into pumping system.

Wharf Case Study No 4

Northfleet, Botany Marshes, Kent

This wharf is one of only three in the U.K. where a pump discharge system is operated from the vessel.

Northfleet wharf from dredger

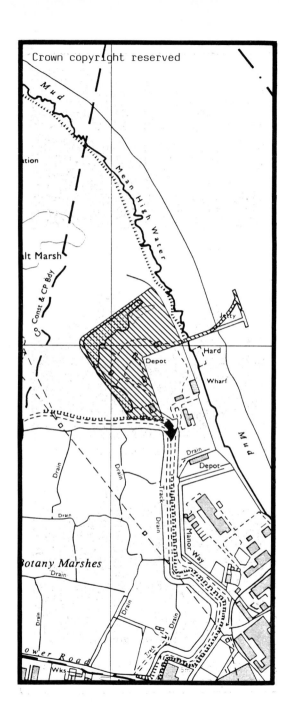

Wharfage/Plant: Washing, grading and crushing plant.
Plant rated at 240 t/hr with a capability of producing in excess of 500,000 tonnes per annum.
Concrete batching plant.

Products : +40mm, 40mm, 20mm, 10mm, 6mm (medium sand).

Stockpiles : unprocessed -7,000t (held in two tanks of 5,000t and 2,000t capacity) plus at least 7,000t held in stockpiles around tanks.

: processed -20mm 11,000t
            10mm 13,000t
            Sand 14,000t

Source of wash water : River Thames slurried cargo is allowed to drain out in settling ponds. Clean water used in processing plant but source to be changed to incorporate a mains water closed circuit washing system.

Silty water settled before sluice gates are opened to release excess water back into the Thames. The discharge is monitored by the P.L.A. for suspended solids and is satisfactory.

Material Quality: Dredged cargoes are of good quality for concreting market. No user prejudice. Sand landed for sub-base work in connection with M25 works south of Swanley.

Market penetration:

No marketing to north via Dartford Tunnel because of RMC operation at Purfleet Wharf (see Case Study No 3).

PLANNING CONSIDERATIONS:

The isolated position of this wharf on marshland does not give rise to any complaints or planning problems. The wharf is located on the end of a private road owned by the Company. Several other industrial users have access to this road.

COMMENT:

This wharf is ideally placed both in terms of its industrial position (on marshland), close to a strategic road network (A2/M20/M25), and to a market area where land won sand and gravel supplies have declined in recent years.

The change in the source of wash water has arisen only because there have been times in the past when during a long dry summer the amount of fresh water flowing down the River Thames has been reduced such that the tidal excursion (and chloride concentrations) can penetrate higher upstream. Although under such conditions the quality of the wash water has been satisfactory, the switch to a mains supply is seen as security in the event of a drought with associated high river chlorides.

Pump discharge into tanks

Processing plant and washwater pump

mv Sand Weaver and controlled discharge from lagoon to River Thames

WHARF: Alpha Jetty, Cliffe, Kent.   **WHARF AREA** 0.2 ha   **NGR** TQ 706767

PLANT: North Sea Terminal.   **PLANT AREA** 5.5 ha   **NGR** TQ 720755

OPERATOR: Brett Marine Aggregates Ltd.

LAND OWNER: Blue Circle Industries PLC.

SITE DESCRIPTION: Wharf located on right (south) bank of River Thames, five miles north east of Gravesend. The jetty which is approximately one mile from processing plant and linked by field conveyor is located on Cliffe Marshes in an area formerly worked for clay and chalk for cement manufacture. The site has a processing plant for dredged material, value-added plant and a rail link.

COMPANY DELIVERING MARINE DREDGED MATERIAL: ARC Marine Ltd (formely Westminster Gravels Ltd, previously Marinex Gravels Ltd).

MATERIALS LANDED: sand and gravel mainly from Norfolk banks.

DATE LANDINGS OF MARINE DREDGED MATERIAL COMMENCED:

    1972 by former Marinex Gravel Ltd.
    1975 by Brett Marine Aggregates.

OWNERS/OPERATORS POLICY TOWARDS WHARF: To maximise land assets of former extraction areas associated with cement manufacture. Cement works now demolished.

OPERATIONAL CONSIDERATIONS:

Berth: Vessel size currently limited to 10,000 dwt.
    Size of vessel normally used : 3,350 dwt.
    Depth of berth water : 6.5 at MLWM.

Wharf Case Study No 5

Alpha Jetty/North Sea Terminal

Cliffe, Kent

This site is reputedly capable of the largest throuhgput of marine sand and gravel in Europe.

It is the furthest wharf downstream of the River Thames located on marshland and is rail-linked.

The processing plant is unusual in that it has stocking in conveyors in addition to the normal stocking out conveyors.

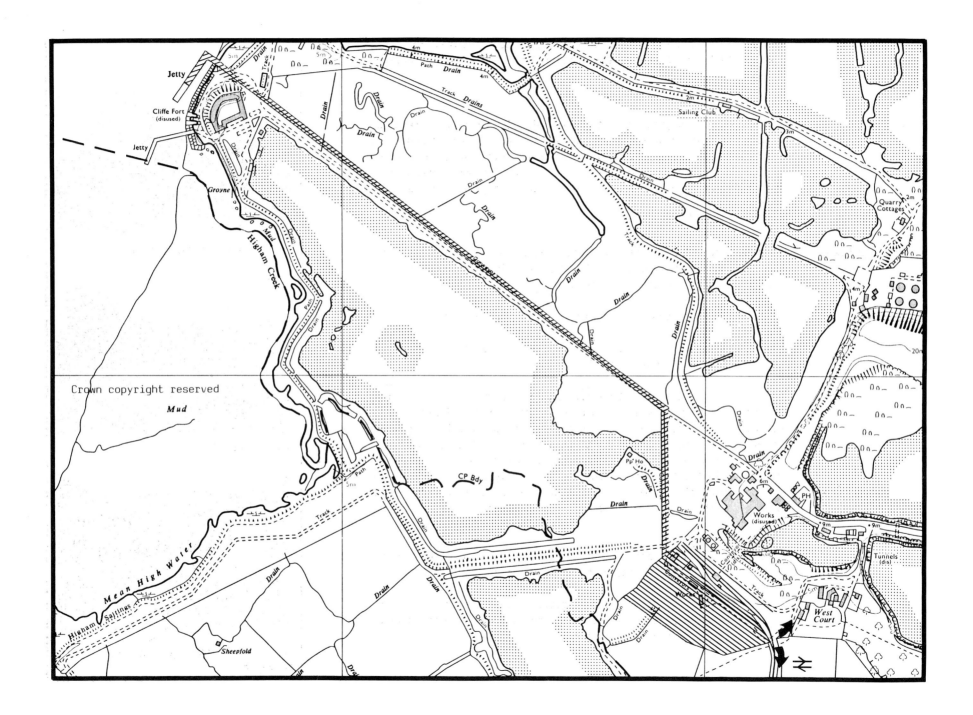

Jetty 180m long and 20 metres offshore is to be reduced in length during 1985 for ease of maintenance but will not restrict current operations.

Discharge: Discharge of cargoes can take place at 1,200/hr at all states of tide. Self-discharge from vessel luffing boom to primary raw stockpile on wharf for onward movement via 2 km long conveyor system over marshes to plant site. Landing and processing of materials can take place 24 hours per day, 7 days per week depending on demand.

Wharfage/Plant: Sand and gravel processed by washing, grading and crushing through a plant rated at 400 tph; 1,500,000 tpa. Linatex sand plant with three towers and dewatering screens.

Products : +40mm, 40mm, 20mm, 10mm and 5mm (medium sand).

Processed material is sold ex-wharf with deliveries by road or rail; there is a concrete batching plant on site.

Rail link into site: two 128 metre sidings. Wagons are loaded @ 6½ tonnes per minute via an overhead travelling tripper fed by conveyor from the main plant. Once overhead storage bins are full material is stocked out and the plant features both stocking out and stocking in conveyors. These are essential as rail loading is not only faster than processing but frequently operated outside of normal production time. Blending of material to specification is done through a data-processing system operated from the main control room.

Source of wash water: former chalk pit now flooded with fresh groundwater. Chloride content has always met specification.

Means of discharge of silt laden water: through a fines reclaimer with water discharged to clay lake. Fines discharged onto conveyor taking processed sand to stockpiles.

General view of plant area

Processing plant

Close up of stocking in and out conveyors

Concrete batching plant

Loading wagons by overhead travelling tripper

Loaded train leaving Cliffe

Stockpiles : 60,000 tonnes (as dredged).
150,000 tonnes (processed).

5½ day week worked at present.

Material Quality: Both chloride and shell content satisfy BS 882: 1983. To achieve the exacting standards required by the construction industry the plant throughput is reduced to around 325 tph. Regular control testing is carried out.

Market penetration: Currently an approximate 70% of total sales are railborne to company owned depots at Salfords (between Redhill and Gatwick) and Purley (near Croydon) as well as to markets in the GLC area and other parts of the South East.

For economy of movement, trains of 18 company owned wagons carrying 1300 tonnes of material are dispatched from Cliffe and divided at Salfords with the remaining half proceeding to Purley. Customers privately owned wagons are dispatched from Cliffe in similar movements (when possible, as some sidings cannot accomodate 18 wagon trains) supplying their own ready mixed concrete plants in the London area. British Rail, in their own hopper type wagons, occasionally collect granular fill material for civil engineering track maintenance.

The remaining percentage of sales are road borne, usually owner driver vehicles, serving principally the Medway towns, including the Group's own ready mixed plants, up to approximately 20 km radius. Competition from land pits and other marine sources at Sittingbourne to the east and Gravesend/Greenhithe to the west normally limit market penetration in those directions.

The Company is part of the Robert Brett Group which have established land-won sand and gravel pits in Kent. Inter company sales ex-Cliffe can assist other parts of the Group when required.

PLANNING CONSIDERATIONS:

Site located: in a rural area remote from housing. Only former mineral workings in area now flooded.

M2 motorway 11 km, road access via mainly 'B' class road. Although there are no restrictions on loading of trains and vehicles, trains leave through the day and night but lorries currently only operate during the daytime. (Lorries pass along a 'B' class road through a small village and objections might be raised to night-time vehicle movements).

COMMENT:

This complex is the largest single marine aggregate wharf/depot in Europe, not only in size but also in the available capacity of tonnage processed per annum. Viability of operations was in serious doubts in early years before present owners took over. The demise of Marinex Gravel was due largely to a combination of three factors:

i) part of the dredging fleet (Marinex V) was poorly designed - it has now been scrapped. There were visions that sewage sludge could be taken out to the North Sea and after dumping, Marinex V would then dredge for sand and gravel. Sludge dumping never materialised.

ii) the reserves of sand and gravel had not been fully evaluated and dredged cargoes were unsatisfactory;

iii) the material was not produced economically.

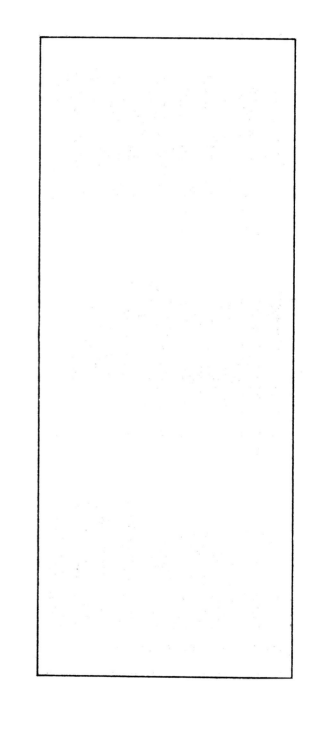

WHARF: Northwall Quay, Dover, Kent **AREA** 0.44 ha **NGR:** TR 319407

OPERATOR: Nickolls Ltd. (land sub-let from Corralls [coal merchants]).

LAND OWNER: Dover Harbour Board

SITE DESCRIPTION: A small wharf with access via the Wellington Swing Bridge immediately adjacent to the Western Hoverport. Lorries run into the site via a narrow slip road between the plant site and the Hoverport.

COMPANY DELIVERING MARINE DREDGED MATERIAL: ARC Marine Ltd.

MATERIALS LANDED: sand and gravel mainly from Owers Banks off Worthing.

DATE LANDINGS OF MARINE DREDGED MATERIAL COMMENCED: 1965

OWNERS/OPERATORS POLICY TOWARDS WHARF: The owner wishes to increase facilities for the Hoverport and to re-locate this operation into the Granville Dock. The operator wishes to maintain a presence in this area despite new constraints at Dover (see Comments at end).

OPERATIONAL CONSIDERATIONS:

Berth: Vessel size limited to 2,500 dwt
Size of vessel normally used : 2,500 dwt
Depth of berth water : 7m (spring maximum)
Length of wharf : 110m
Tidal Harbour. 2,500 dwt vessels can only berth on days with spring tides > 6m.
Small vessels of 700 dwt can berth on any day but this involves a grab discharge. The operator finds that small cargoes are becoming increasingly uneconomical and therefore prefers the larger cargoes with restricted berthing times.

Wharf Case Study No 6

Northwall Quay, Dover, Kent

This wharf is alongside Dover International Hoverport and is now closed due to pressure to extend car parking facilities for tourists.

The operations have been re-located in Dover Harbour onto a much smaller site. Road access to the site involved crossing a swing bridge which is opened for the passage of vessels including dredgers for another marine sand and gravel operation.

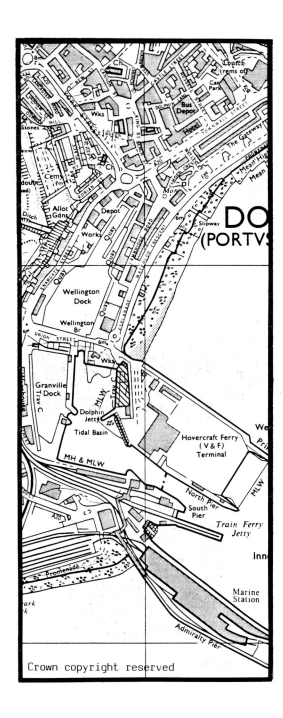

Discharge: Self discharge to wharf stocking bays within 2½ hrs. Clam-shell grab on wharf moves material into plant for processing.

Wharfage/Plant: Washing and grading plant but no crusher. Plant rated at about 60 tph., 100,000 tpa. Concrete batching plant. The operator has not re-furbished the appearance of the plant (cladding etc.) because of an uncertain future at Northwall (see below).

Products : +40mm, 40mm, 20mm, 10mm, 5mm, coarse sand.

Some material sold 'BAD' for fill.

Source of wash water : mains. Silt taken off site for disposal.

Stockpiles : 4-5,000 tonnes (as dredged) and
             1,000 tonnes (processed) equivalent
               to 1-2 days supply.

No restriction on hours of working or discharge of cargo but currently a 5½ day week is worked.

The Wellington Swing Bridge is normally opened about once a day and causes delays for lorries of between 15-30 minutes.

Material Quality: Processed material always meets chloride and shell limitations in BS specifications. Some reject material because of a lack of space to install a crusher but marketing to date of this has not prevented any problems.

Market penetration:

Up to 15 radial km to north west and north and 10 radial km to south west before competition from land- won sources. Access through Dover is generally good to markets beyond the town but conjestion can arise at peak holiday times with tourists.

PLANNING CONSIDERATIONS:

Location of site is in an area where there are no gravel pits. Development of Union Quay (see below) has required planning permission and difficulties arose with the first application because of land required by the Department of Transport for road improvements. Although that application was permitted a second application was also submitted and approved for a revised lay-out of plant and access arrangements.

COMMENT:

At the time of the time when this case study was prepared (December 1984) the Harbour Board was in discussion with the operator to move their point of landing by 31 May 1985 to the Union Quay (96m long) in Granville Dock. Union Quay was formerly used for coal and stone. Although that quay is in a locked basin this does not present operational problems as dredgers of 2,500 dwt are already limited to high spring-tides to berth at Northwall. Also, the Union Quay avoids the need to use the Wellington Swing Bridge by the operator's vehicles.

The re-located wharf is only one-third the size of Northwall Quay and because of this all processing plant and stockpiles will have to be elevated in height. Although the new plant will have the same production capability, the operator will be constrained and the anticipated output may be reduced by some 50% due to a shortage of stock-piling areas.

Wharf adjoining hoverport

Entrance to plant area

Processing plant

CASE STUDY No.7 A

WHARF: North Quay, Newhaven   **AREA** 1.62 ha   **NGR** TQ 447007

OPERATOR: ARC (South Eastern) Ltd.

LAND OWNER: -

SITE DESCRIPTION: A wharf on the left (east) bank of the River Ouse located within an industrial area with another marine sand and gravel wharf to the north and a timber yard to the south. Small builder's yard operated by ARC in north east corner of site. Good access to A 259. On the right (west) bank of the River Ouse is a light industrial area on a site known as The Island. The wharf lies upstream of Newhaven Swing Bridge.

A Case Study on this wharf was agreed by the Steering Committee (Section 1.1) for inclusion in this report. However, when a draft of this Case Study was circulated to that Committee the Company affected decided to withhold publication of further details.

Wharf Case Studies No 7 A and B

North Quay, Newhaven, East Sussex

Two wharves are described here which adjoin each other. One is rail linked. The narrowness of the channel and size of vessels normally used necessitates careful manoeuvring to berth and sail. A swing bridge is also involved.

North Quay and River Ouse

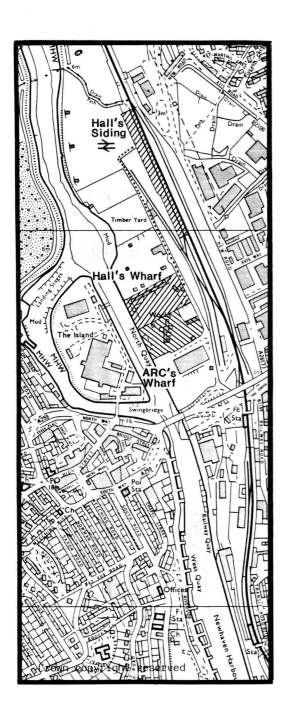

CASE STUDY No.7 B

WHARF: North Quay, Newhaven **AREA** 1.57 ha **NGR** TQ 447008

RAIL SIDING 360m long. **NGR** TQ 447012

OPERATOR: Hall Aggregates (South East) Ltd.

LAND OWNERS: (Wharf) British Ferries
(Sidings) British Rail

SITE DESCRIPTION: A rail linked wharf located on the left (east) bank of the River Ouse in an industrial area with a scrap metal yard to the north and another marine sand and gravel wharf to the south. Good access to A 259. On the right (west) bank of River Ouse is a light industrial area on a site known as The Island. The site lies upstream of Newhaven Swing Bridge.

COMPANY DELIVERING MARINE DREDGED MATERIAL: South Coast Shipping

MATERIALS LANDED: Sand and gravel mainly from Owers Bank and some from Pot Bank off Isle of Wight.

DATE LANDINGS OF MARINE DREDGED MATERIAL COMMENCED: 1973

OWNERS/OPERATORS POLICY TOWARDS WHARF: Although the owners have no firm policy towards the wharf being a recent acquisition from Sealink, the operator's policy is to maximize the area for aggregate development.

OPERATIONAL CONSIDERATIONS:

Berth: 95 metres long.

North Quay

Pedestal grab crane

Concrete batching plant

View of sidings from plant site

Loading wagons in siding

Normally 1,500 dwt vessels berth up to one hour before high tide (6m tide). The maximum size vessel is 2,000 dwt but these can only berth up to ½ hour before high tide given a tide greater than 6 m. 3 to 4 vessels per week.

Discharge: Grab discharge from pedestal-mounted crane with 4 yd³ clam shells. 1,500t can be off-loaded in 7-8 hours.

Wharfage/Plant Washing, grading and crushing plant rated at 150 t/hr., approximately 300,000 tpa.
All oversize material (> 40mm) is crushed. Concrete batching plant. Final product hopper has three discharge shoots : i) to the concrete batching plant, ii) to road vehicles and iii) not used but was originally planned as a conveyor line to rail wagons.

Products : 40mm, 20mm, 10mm, 5mm mainly medium sand, BAD.

Source of wash water : mains. Regulated and tested daily for chloride concentration. Closed-circuit system which is topped up each day after water is drained down from the plant and discharged to the River Ouse. Silt is removed for off site disposal at a landfill.

Stockpiles : 3,500 tonnes (unprocessed).
25,000 tonnes (processed).

(6,000t sand; 7,000t @ 10 mm; 10,000t @ 20 mm; and 2,000t @ 40 mm).

Single shift worked (10 hour day) 5½ days/week. Wharf has operated 24 hrs/day in the past to meet sudden demands.

Material Quality: Cargoes landed are mainly as dredged i.e. no on-board screening. Screens used for specific sand cargoes. User prejudice is not a problem.

Market penetration: Over 50% of output goes by rail to Crawley and Tolworth.

Material is taken from plant site by 20t dump trucks along a 500 metre private road to a rail siding. Material is stacked in separate bays : sand, 10 and 20 mm (40 mm only rarely). The bays have capacity to fill two trains (i.e 1,600 tonnes). Loading is done by front end loader with a 4 yd^3 bucket. All aggregate is weighed automatically in the bucket before discharge to the wagon. A train is loaded in 2 hours. Five trains leave per week: four to Crawley at 09.25 am and one to Tolworth at 20.26. Trains only supply concrete batching plant at present. Trains are loaded throughout the day as required. Twelve wagon trains, each wagon carrying 65 tonnes of aggregate.

Little competition from land-based pits because of lack of pits or pits just producing sand.

PLANNING CONSIDERATIONS:

No restrictions on hours of working.

COMMENT: Normally South Coast Shipping will berth first at North Quay, turning the vessel in The Pool (at north end of The Island) so that it docks on the port side. ARC Marine with their larger dredgers will berth to starboard. South Coast Shipping will sail first, thus allowing ARC manoeuvring space to turn in The Pool.

Siding: note storage bays on right

WHARF: River Road, Littlehampton   **AREA** 0.18 ha   **NGR:** TQ 024020

OPERATOR: ARC (South Eastern) Ltd.

LAND OWNER: -

SITE DESCRIPTION: A small site on the left (east) bank of the River Arun. River Road is narrow with mixed development: residential properties directly opposite site entrance. River Road is one way (east bound) immediately west of wharf. Close to A259. On the opposite river bank are holiday moorings / boat yard with a golf course beyond but generally screened from view by flood defences and vegetation.

A Case Study on this wharf was agreed by the Steering Committee (Section 1.1) for inclusion in this report. However, when a draft of this Case Study was circulated to that Committee the Company affected decided to withhold publication of further details.

COMMENT: This is one of two wharves in Littlehampton, the other being operated by Tarmac. West Sussex is largely dependent on landings of marine aggregates and therefore competition is mainly from other marine landings rather than land-won gravel.

Wharf Case Study No 8

River Road, Littlehampton

West Sussex

This is a small wharf located in an area of mixed development with housing directly opposite the site entrance.

WHARF: Kendalls Wharf.   **WHARF AREA** 2.27 ha.   **NGR** SU 674033

OPERATOR: Kendall Bros. (Portsmouth) Ltd.

LAND OWNER: Kendall Bros. (Portsmouth) Ltd.

SITE DESCRIPTION: Wharf built out onto saltings along western shore of Langstone Harbour which is an SSSI. Shallow approach channels. Located to the east of Portsmouth by the A2030.

COMPANY DELIVERING MARINE DREDGED MATERIAL: Kendall Bros.(mv Kaibeyar & KB,864 and 432 tonnes cargo respectively).

MATERIALS LANDED: Sand and gravel from mouths of Langstone Harbour and Portsmouth Harbour (includes dredging from Small Sands within the SSSI and for which local authority planning permission exists).

DATE LANDINGS OF MARINE DREDGED MATERIAL COMMENCED:

     Present wharf 1962.
     Firm started at Velder Lake 2 km south of present site in 1898.

OWNERS/OPERATORS POLICY TOWARDS WHARF: Small scale plant improvements foreseen. Otherwise to pursue steady trade.

OPERATIONAL CONSIDERATIONS:

Berth: Approximately 100m in length, in main channel; vessels ground at low water. Piled steel/concrete wharf front. Mooring for two small vessels facing upstream - turn on warp to quay on departure. Accessible 3.5 hours before, to 2 hours after, high water.

Wharf Case Study No 9

Kendalls Wharf, Portsmouth, Hampshire

Wharf located within a Site of Special Scientific Interest (Langstone Harbour) and built out over saltings to narrow channel. There is pressure for increasing recreational use of the area.

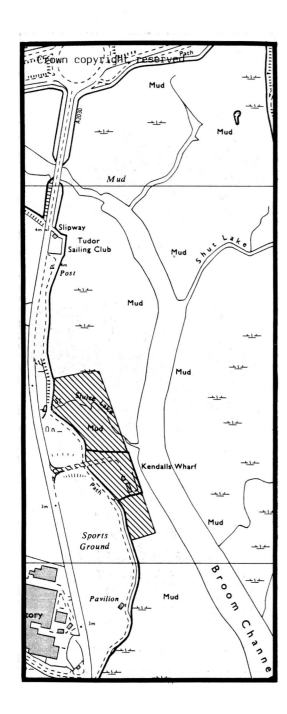

Discharge: Vessels discharge on berthing. Crawler mounted grabs(2), mv Kaibeyar taking 4½ hours and mv KB taking 2½ hours to unload. Discharge initially feeds process-supply hoppers, then stockpiles on quay. Cranes used to feed hoppers once discharge completed.

Wharfage/Plant:

1) Dry screening @ 20mm, 'all in' material passing stored in bin and sold - 50% of all sales.

2) Wet screening @ 70 tonnes per hour.

Products : 40mm, 20mm, 10mm, 5mm and 'all in' material stored in timber pens. No crushing - oversize material sold for site fill. Moved with front-end loader.

3) Cargo also sold as dredged - no stockpile kept but customers requirements met directly. "Sea sand"-small quantities dredged from estuary and sold for recreational purposes (riding school floors, sand pits etc.).

Source of wash water : mains. Some recirculation, silt trapped and dumped in the local tip.

Stockpiles : 17,000 tonnes (unprocessed and + 40mm)
             11,000 tonnes (processed)

Plenty of room but no need to increase stockpiles.

Material Quality: Chloride - satisfactory. Shells (blackened oyster shells) occasionally landed, especially when dredging from Spit Sands: largely removed by 20mm screen.

Market penetration: 25 km, using own haulage plus that of customers.

PLANNING CONSIDERATIONS:

Conflict with local authority as latter want to develop Langstone as a recreational harbour. Local authority would like to see wharf closed but lack of an alternative, suitable site and lack of finance makes closure unlikely. It is the County Planning Authority's policy to restrict development here that would increase disturbance or visual intrusion: difficulty in obtaining planning permission for new garage on site. No restrictions on hours of working.

Close to good road network (M27/A27/A3(M)).

COMMENT:

Kendalls do not dredge in SSSI with the exception of land-based berth deepening.

The presence of this wharf acts in a positive way to reducing the number of moorings along this length of the harbour, thus allowing more maneouvring space for the County Council's school of sailing instruction.

Tarmac wharf (Bedhampton) also in Langstone Harbour. Tarmac have dredged a channel to their wharf.

mv Kaibeyar at Kendalls Wharf

Wet screening plant and sand plant

Dry screening plant and sand plant

WHARF: Upper Quay, Fareham.   **WHARF AREA** 0.42 ha.   **NGR** SU 581058

OPERATOR: Northwood (Fareham) Ltd.

LAND OWNER: Northwood (Fareham) Ltd.

SITE DESCRIPTION: A small wharf on the right (north) bank of Fareham Lake, a tidal creek at the head of Portsmouth Harbour. The northern part of Portsmouth Harbour, including Fareham Creek, is in the Portsmouth Harbour SSSI. Site adjoins A27 Fareham By-Pass by a roundabout to the north. Immediately adjoining the wharf are various light industrial enterprises including a boatbuilder. Dominating the sky-line to the north (on the far side of the railway viaduct) is Brickhouse Dudley's foundary. Housing occurs to west of site by Town Quay and boat yards to south west. Both the Quay and the housing lie in the Town Quay Conservation Area. Recreation ground to east. Open fields to south on opposite bank.

COMPANY DELIVERING MARINE DREDGED MATERIAL: Northwoods (Fareham) Ltd.

MATERIALS LANDED: Sand and gravel mainly from Bembridge Harbour. Some from Hamilton Bank and Solent Bank.

DATE LANDINGS OF MARINE DREDGED MATERIAL COMMENCED: April 1984 (re-located from Bath Lane Quay ½ km upstream to the east). Wharf previously used for coal deliveries and for a time, marine sand and gravel - see Comment below.

OWNERS/OPERATORS POLICY TOWARDS WHARF: To make maximum use of the space available. Part temporarily leased to Hampshire County Surveyor as a storage compound for road signs etc.

Wharf Case Study No 10

Upper Quay, Fareham, Hampshire

A small site opened in 1984 following re-location by the owner from an established wharf ½km away. The Quay, which is within a Conservation Area, is probably unique in that berthing can take place on any one of three sides.

Sand and gravel only dry-screened before sale at present.

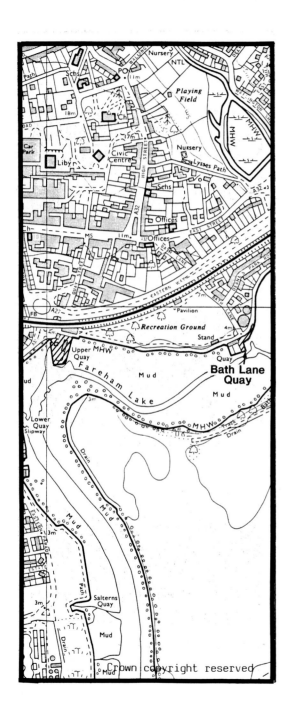

OPERATIONAL CONSIDERATIONS:

Berth: Vessels can berth on one of three sides to the Quay. The depth of water on the east side is one foot shallower. With shallow draft vessels two ships per tide would be considered a maximum.

Discharge: Mobile grab (38 RB) with $1\frac{1}{2}m^3$ clam shell. (approx $3\frac{1}{2}$ tonnes).

Wharfage/Plant: Mobile hopper with elevating conveyor to screen off rejects. Separate crawler grab 19RB (about one $3/4m^3$ bucket) for loading lorries.

Products :   3/4" all in   (<20mm screened)
               Rejects   (>20mm including shells)

Stockpiles : currently @ 2,000 tonnes (as dredged)
                     250 tonnes (screened)

Material Quality: No problems with chlorides. Dredged material from the Hamilton Bank is especially rich with oyster shells but these are screened before sale. Cannon balls were a problem in the past but rarely nowadays since these are found mainly in the surface layers of the banks.

Market penetration: Company do not have their own vehicles and rely solely on local builders etc to collect material.

Deliveries : 15 km to west; 35 km to east; and 25 km to north west.

PLANNING CONSIDERATIONS:

The operation of this wharf for the landing of sand and gravel did not require planning permission, the use was already established. This new location for the operator is better situated in terms of access and avoidance of predominately residential areas. However, it is within a conservation area and adjacent to an SSSI. Good access to A27.

No restriction on hours of working : cargo discharges are mainly confined to daytime. Local people are concerned about the noise and visual intrusion of the aggregates wharf. The report of the Inspector into the Hampshire Minerals Local Plan recommended, in January 1986, that this site should be included within the Council's policy where development would not be permitted which would result in increased disturbance or visual intrusion.

COMMENT: A small wharf re-located for operational reasons. The original Bath Lane Quay involved the passage of lorries through a residential area to the A27. That site is currently the subject of a planning application for residential development in preferance to continued commercial use.

Prior to 1984 Upper Quay was owned by Powell Duffryn who used it for unloading and processing coal. During the period 1968-78 marine sand and gravel was also landed (by South Coast Shipping). The material was washed and screened.

A mobile washing plant has been proposed by the current operators to enable an increase in output from the wharf but the County Planning Authority resolved, in July 1985, to issue an enforcement notice and stop notice if an aggregate processing plant was brought into operation.

Although it would be theoretically possible to increase the daily throughput to a maximum of 2,000 tonnes using 4 vessels (two tides x 2 ships), the planning constraints on this site seem likely to preclude this realisation.

General view of wharf

19RB used for loading lorries

Mobile screening equipment

WHARF: Baker's, Millbank Street, **WHARF AREA** 0.52 ha   **NGR** SU 434124
Northam, Southampton, Hants.

OPERATOR: Pioneer Aggregates (UK) Ltd.

LAND OWNER: Bees Finance.

SITE DESCRIPTION: A small wharf on the right (west) bank of the River Itchen. To the north is a yachting complex and to the south is a vehicle container storage area which shares the same narrow entrance to the wharf. Opposite the site entrance are multi-storey residential flats. On the opposite side of the river are commercial operations, including another wharf (Willments Yard) used for marine landings.

COMPANY DELIVERING MARINE DREDGED MATERIAL: Northwood (Fareham) Ltd.

MATERIALS LANDED: Sand and gravel mainly from Nab Bank and some from Pot Bank, off the Isle of Wight.

DATE LANDINGS OF MARINE DREDGED MATERIAL COMMENCED: 1976.

OWNERS/OPERATORS POLICY TOWARDS WHARF: To maximise available space.

OPERATIONAL CONSIDERATIONS:

Berth: Vessels normally berth at high water to make manoeuvring easier because of immediate proximity of marina on upstream side of wharf. (At one time the marine jetties extended illegally across part of the wharf frontage). Care has to be taken not to approach the berth too fast or being drawn along on a flood tide. Berthing can be up to 3 hours before high water if necessary, especially if 2 vessels are to berth on the same tide.

Wharf Case Study No 11

Baker's Wharf, Southampton, Hampshire

A small conjested wharf with a narrow access to the public highway. Because of nearby housing only the crane may be used at night time for off loading cargoes.

Operator runs a second wharf nearby, now only for storage of processed materials.

Grab crane and raw stocks

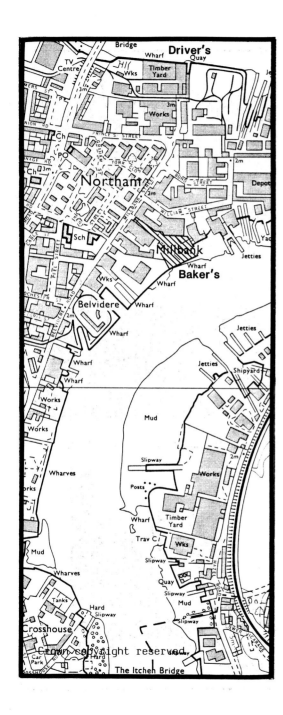

| Discharge: | NCK crawler grab crane with $3\frac{1}{2}m^3$ clam shells (5 tonnes). Discharge completed in 1 3/4 - 2 hours. Material is stockpiled for 12 hours then fed onto a surge pile with conveyor tunnel beneath. Metal detector in tunnel over conveyor with klaxon alarm should metallic objects pass detector. |

Wharfage/Plant: Washing, screening and crushing plant. Plant rated at 150t/hr with a projected maximum annual throughput of 300,000 tonnes. Current production rates are around 800t/day and approximately 200,000 tonnes per annum. Concrete batching plant.

Source of wash water: mains. Water re-cycled and topped up as necessary. Flocculent added to silt laden water. Two settlement pits. One used for settling silt while the other is drying out in readiness for disposal to local tip by lorry.

Products : +40mm, 40mm, 20mm, 10mm and 5mm (medium sand), and BAD.

Stockpiles :   2,500t (unprocessed)   (
                                                          (see Comment below.
              6,000t (processed)   (

Material quality: Little over-size material despite landing of as dredged sand and gravel i.e. with no on-board screening. No problems with stone/sand ratio, salt or shell contents. Daily checks on chloride concentration: water sampled at 17.00 hours and the stocked material at 06.00 hours.

On rare occasions ammunition shells are landed. Timber (as small black pellets) is sometimes a problem, mainly from Pot Bank cargoes. These are caught on the 10mm screens in the washing plant but where they have passed through into the end product in concrete they shrink and pop out. This is visually unacceptable to some customers but does not affect the overall strength of the concrete. Whiteweed is also an occasional problem as the remenants of the seaweed adhere to the stone fraction.

User-prejudice is not a problem.

Market penetration: 60% of material goes to company's concrete batching plants. Marketing about 15 km to east, 20 km to north and 30 km to south-west.

PLANNING CONSIDERATIONS:

No restriction on hours of working although vessels do not generally berth late at night and the company operate a 5½ day week at present.

Rubber matting fixed to grab to reduce any noise problem following comments from flat residents near the site entrance.

The access is about 100 metres long and is narrow with lorry trailers parked on one side. At its junction with Millbank Street is a warehouse for timber where vehicle loading partially obstructs the free flow of vehicles into and out of the wharf.

COMMENT:

This wharf is used in conjuction with Drivers Wharf (½km to north) operated by the same company. Previously the latter site, with a storage capacity of 11,000 tonnes, had been used soley for marketing as dredged cargoes mainly for fill in motorway construction and for re-loading onto a smaller dredger (Sea Driver) to take material to West Medina Wharf, Newport, I.O.W. During 1985 landings at Drivers Wharf ceased and that site is now used for stocking processed material from Bakers Wharf.

With additional space at Drivers Wharf, there is some scope for an increased throughput at Bakers Wharf.

Concrete batching plant

Processing plant and stocking bays

Site access

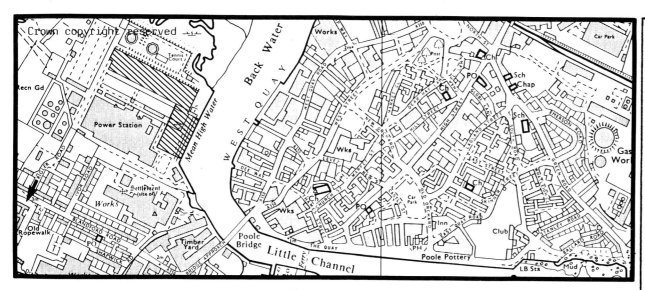

WHARF: CEGB Power Station, Poole   **WHARF AREA** 2.13 ha   **NGR** SZ 003906

PROPOSED OPERATOR   Pioneer Concrete (UK) Ltd

LAND OWNER   CEGB

SITE DESCRIPTION: A flat L-shaped area within the grounds of CEGB Power Station and two areas within the gardens of 116 and 118 Blandford Road at their junction with Rigler Road. To the west is an electricity sub-station and screened from residential development to the south by the power station and other industrial buildings. To the north, beyond CEGB oil tanks, are mud-flats and industrial quays face the site across Back Water Channel to the east.

COMPANY WHO PROPOSED TO DELIVER MARINE DREDGED MATERIALS:

     Westminster Gravels Ltd   (now part of ARC Marine Ltd).

Wharf Case Study No 12

CEGB Site at Poole, Dorset

This site is included in the case studies as an example of an approval which has not been implemented. The development was the subject of a public inquiry in January 1982. The Inspector, who determined the appeal under delegated powers in March 1982 (ref: T/APP/5189/A/81/10700/G3), dismissed that part of the proposals relating to junction improvements and allowed the appeal relating to the use of the CEGB's land for the importation and processing of marine aggregates but limited the use of the site for this purpose to five years from 25 December 1982.

The appellants found the decision unsatisfactory and therefore did not proceed.

This case study follows a similar format to the others for ease of comparison and is based solely on the Inspector's decision letter.

CEGB site from Poole Bridge

MATERIALS TO BE LANDED: sand and gravel from off the Isle of Wight.

DATE LANDINGS OF MARINE DREDGED MATERIAL WERE TO COMMENCE: 25 December 1982

OWNERS/OPERATORS POLICY TOWARDS WHARF: Both CEGB and the Poole Harbour Commissioners gave strong support for this development which was a temporary planning permission for the use of the site only for 10 years. A detailed application for the processing plant and structure was to follow. The lease was to run from 25 December 1982 terminating at any time after 3 years subject to 2 years notice, thus permitting a mimimum occupation of 5 years.

OPERATIONAL CONSIDERATIONS:

Berth: 109.7 metres long. Access for vessels via Little Channel and Poole Bridge, a lifting bridge carrying the A 350. Proposed that an 800 dwt dredger would be chiefly used involving 10 trips per week through Poole Bridge, although fewer additional openings than this would be required because use would have been made of regular bridge opening times whenever possible. Only 2 or 3 additional openings per week would have been likely to occur. Traffic conjestion, outside peak hours, was not a problem. A larger dredger could be used involving less trips through the bridge.

Discharge: Not stated.

Wharfage plant: Processing plant for washing and grading aggregates rated with a throughput of 200,000 tonnes per annum.

Source of washwater : not stated.

Stockpiles : not stated.

Material Quality: Material to be supplied was of a quality suitable for use in concrete manufacture.

Market penetration: 40 lorry loads per day each carrying 20 tonnes (ie 80 lorry movements per day) to serve Poole and its hinterland where a potential growth existed. About half of the output was to supply concreting batching plants at Sturminster Marshall and Ferndown (some 11 km away).

PLANNING CONSIDERATIONS:

The Cases: It is Government and regional policy to increase supplies of marine dredged aggregates and Poole is the only port in the county which could be used for this purpose. The S E Dorset Structure Plan, as modified by the Secretary of State, encourages further development of the commercial port of Poole. An existing quay (Wessex Wharf) downstream of Poole Bridge receives marine sand and gravel but its potential spare capacity (if any) was in dispute. Land-based aggregates were also available.

Lorry movements were not to be very different to that which existed until 20 years previously when fly ash was taken off site before conversion of the power station to oil. However, an overall increase in all trafic had been recorded along Blandford Road in the previous ten years.

The two main impediments to the development were:
    i) the Blandford Road/Rigler Road junction, and
    ii) Poole Bridge.

The Inspector accepted the views of the planning authority and Poole Borough Council that the proposed junction improvement aimed at overcoming a highway objection to the increased use of the existing junction (involving about $18\frac{1}{2}$ m² from each of two gardens to provide safe turning for vehicles) would have a seriously harmful effect on the amenity of residents of the corner houses. The Borough Council also owned the houses and were unwilling to sell the land. The positive features about the junction were its capacity and good visibility.

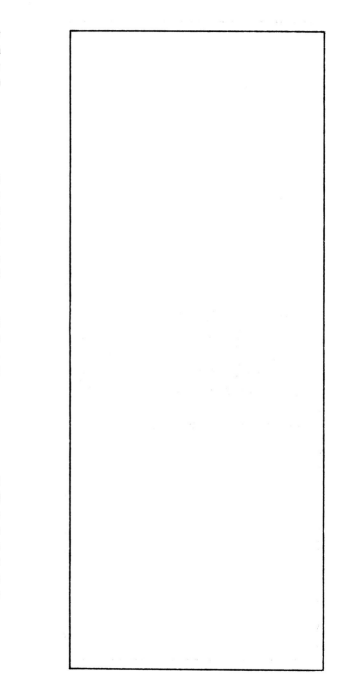

Rigler Road/Blandford Road junction

A site demonstration was laid on with 3 and 4 axle lorries to show their turning capabilities at the junction. The Inspector was not satisfied that the shortcomings with the existing junction were sufficient alone to prevent the proposed development.

Although the appellants stated that they could regulate the passage of the dredger to minimise the impact upon traffic crossing Poole Bridge, it was pointed out that commercial vessels have a right of way through the bridge on demand. The Inspector was of the opinion that only experience over a period of time would show whether or not this would add unacceptably to traffic conjestion.

Inspector's conclusions:

* That the planning authority should have the opportunity to review the situation in respect to both the Blandford Road /Rigler Road junction and Poole Bridge after a reasonable period and that a temporary permission for 10 years would be excessive.

* He was mindful of the capital expenditure to be incurred on the project and believed that the 5 years minimum period for site occupation was an appropriate duration for a temporary permission.

* Lorries generated by the development spread throughout the working day would have a generally mariginal effect upon residential amenity in Blandford Road.

Inspector's decision:

* <u>Dismissed</u> that part of the appeal relating to the proposed improvement of the junction of Blandford Road / Rigler Road, and

* <u>Allowed</u> the appeal and granted planning permission for the importation and processing of marine aggregates subject to four conditions :

    1. the development was to commence within five years (ie by 29 March 1987).

2. hours of working limited to 07.30-17.30 Mondays to Fridays and 07.30-12.00 on Saturdays for processing of aggregates and their removal from site.

3. vehicles leaving the site with aggregates not to exceed 8 m in length and have no more than 3 axles except with the consent of the local planning authority, and

4. the use permitted to be discontinued at or before the expiration of 5 years from 25.12.82.

COMMENT: The operator did not proceed with this development for two reasons:

i) Condition 4 did not provide for sufficient security of tenure to merit some £0.75M on investment with plant etc at the wharf, and

ii) following a detailed re-appraisal of the potential market from Poole there was insufficient justification to establish an operation, additional to Wessex Wharf, which could be commercially viable in the light of current mineral policies by both Dorset and Hampshire County Councils towards meeting demand in this area. The main source of material is from Ringwood which had also supplied the Southampton market. Encouragement to increase landings in Southampton has lead to a surplus of land won material being available in the Ringwood area to supply the Poole market. No encouragement has been given to increase landings at Poole.

The operator quite rightly decided to secure an investment elsewhere, in the South East, where the medium and long term supplies of land won aggregates were far less secure.

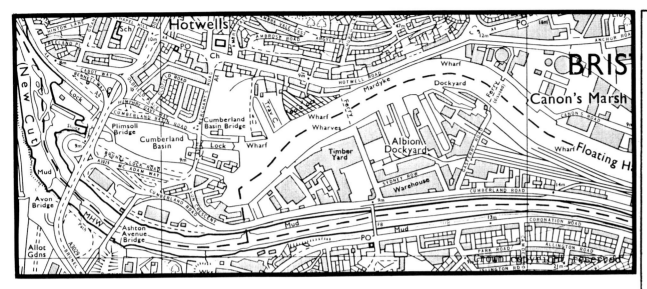

Wharf Case Study No 13

Pooles Wharf, Bristol, Avon

This wharf is in Bristol's Floating Harbour. Dredgers have to pass through a lock and two swing bridges. It was re-located some years ago from nearer the city centre.

Planning restrictions have in one instance precluded value-added plant on a sand wharf in Bristol.

WHARF: Pooles, Hotwell Rd Bristol **WHARF AREA** 0.39ha **NGR** ST 574724

OPERATOR: Bristol Sand & Gravel Co Ltd.

LAND OWNER: British Dredging PLC.

SITE DESCRIPTION: Wharf located in the Floating Harbour, a locked basin. Dredgers have to pass through two swing bridges to berth. Immediately west of the site is another marine sand operator. The area generally is residential with an emphasis on recreational uses in the Harbour. Almost opposite this site is the Baltic Wharf Water Sports Centre.

COMPANY DELIVERING MARINE DREDGED MATERIAL: Holms Sand & Gravel Co and British Dredging PLC.

General view of Hotwells, Bristol

mv Harry Brown at Pooles Wharf

MATERIALS LANDED: Sand from Flat Holm area and Portishead area.

DATE LANDINGS OF MARINE DREDGED MATERIAL COMMENCED: Bristol Sand & Gravel Company Ltd., commenced operations at Pooles Wharf in 1968. However there have been marine landings of sand in Bristol since the early 1920's.

OWNERS/OPERATORS POLICY TOWARDS WHARF: To make full use of the limited space available.

OPERATIONAL CONSIDERATIONS:

Berth: Vessels are not allowed through the swing bridges between 07.45 - 09.00 and 16.45 - 18.00 to minimise inconvenience to road users in the rush hour.

Vessels first enter the Brunel Lock from the River Avon and then pass through the Plimsoll Swing Bridge and into the Cumberland Basin. Traffic meanwhile is diverted over the Junction Lock Swing Bridge. When the Plimsoll Bridge is re-opened to traffic the Junction Lock Bridge is closed to allow vessels to pass into the Floating Harbour. Vessels take about 30 minutes to pass through the lock and two swing bridges.

On Spring tides stop gates below the Junction Lock Swing Bridge are closed to prevent flooding to Bristol City. When this happens further delays to the passage of dredgers can occur with a vessel caught in the Cumberland Basin for up to 3 hours.

Fog in the Avon Gorge or on Horseshoe Bend of the river may prevent sailing or berthing of vessels.

Discharge: Mobile track-mounted grab crane with $4\frac{1}{2}m^3$ clamp shells. Three hours to discharge 700 tonnes. The crane moves along the wharf to feed three holding pens with sands of different quality. When these are full, overhead storage hoppers are filled.

Wharfage: No processing plant. All material screened at sea. Wharf stockpiles can hold 6,500 tonnes equivalent to around 10 days supply. Annual throughput around 0.25 M tonnes. A workshop, weighbridge and site office.

Material Quality: Dredged material is a fine aggregate < 5mm and consists of building sand and fine concreting aggregate. Three materials are dredged, and a blended version allows 4 products to be sold.

Dredging in the Bristol Channel has found little coarse aggregate : non commercially attractive to dredge and market.

No shell problems. Chloride content satisfies B.S.

Market penetration: Although most of the material is sold locally (up to 20 radial km) there is a large geographical market for the fine clean sand for use in school playing fields, jumping pits and bowling greens. This is generally seasonal (in the winter months) and vehicles deliver to Birmingham (via M5) and London (via M4).

PLANNING CONSIDERATIONS:

This site was originally a coal wharf used by the British Dredging Co and has the benefit of an established use.

The wharf came into operation for sand following pressure on Holms Sand and Gravel to move form their original wharf (NGR ST 588723) which:

a) was located nearer to the city centre;

b) was in an area planned mainly for housing and recreational use;

Discharging into storage bins

Stocking pens/storage bins

c) involved the passage of dredgers through a third swing bridge (Prince Street Bridge) where the City Council wanted to minimise the cost involved in operating the bridge; and

d) involved an extra 30 minutes steaming time.

Within the Bristol Docks no further commercial development is encouraged as the City Council wish to turn the area into a recreational waterway. Following from this, value-added plant on wharves is not encouraged.

Housing is proposed opposite this site adjoining Baltic wharf.

COMMENT: Two vessels per day could berth i.e. on each tide. There is competition from land-won sand produced in Gloucester, Dorset and Wiltshire.

Although the site is small, with an internal one way traffic system for lorries, there is a fair access onto the local road network and Motorways M4, M5 and M32.

An adjoining wharf (Hotwells) was refused planning permission on appeal for value added plant (mortar plant).

WHARF: Great Western Wharf, Newport.   **WHARF AREA** 1.5 ha   **NGR** ST 323873

OPERATOR: Tarmac Roadstone Ltd (Western). Pioneer - operate concrete batching plant.

LAND OWNER: Tarmac lease from local landowner.

SITE DESCRIPTION: East bank of River Usk, downstream of George Street bridge. Large tidal range (12m springs). Site surrounded by belt of rough land with no buildings.

COMPANY DELIVERING MARINE DREDGED MATERIAL: Tarmac (Marine) Ltd.
Hoveringham I normally (900 tonne cargo)
Hoveringham IV & V occasionally (1000 tonne cargo)

MATERIALS LANDED: Sand.

DATE LANDINGS OF MARINE DREDGED MATERIAL COMMENCED: 1963

OWNERS/OPERATORS POLICY TOWARDS WHARF: Plant/jetty have just been refurbished. No plans to alter present system.

Wharf Case Study No 14

Great Western Wharf, Newport, Gwent.

Wharf only accessible for approximately 2½ hours over high water. Vessels cannot lie over tide due to foul berth.

Vessels must leave not later than ½ hour after high water due to strong currents even if cargo is not fully discharged.

General view of wharf

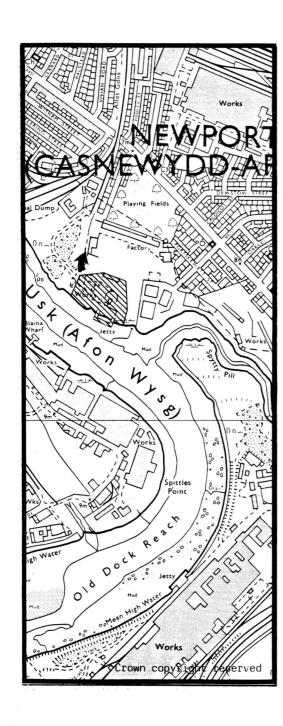

OPERATIONAL CONSIDERATIONS:

Berth: Three free-standing piled structures, two linked by walkway and jetty to shore, third supports hopper at end conveyor. Berthing alongside possible + 2½ hours high water. Ship has to leave shortly after high water otherwise strong currents pin her to jetty. Foul berth, ship cannot stay over low water.

Discharge: Self discharge - scraper system on ship via luffing boom on port bow to hopper

Wharfage/Plant: Two stage (one movable) conveyor system carries sand to stockpile.

Stockpile : 18,000 tonnes (maximum).

Material Quality: Screened at sea to remove coarse material (including whole shell). Builders sand and concreting sand landed. Builders sand better quality than pit sand.

Market penetration: Sand delivered up to normal radius of 40 km.

PLANNING CONSIDERATIONS:

No problems regarding adjacent areas.
Long established site.

COMMENT: Five wharves in Newport

Note : Problem of spillage from conveyor ex ship - shallow water at ship's head, ship is aground for first hour of loading.

: berthing problem due to fast river. Ship steams up to just above wharf, put nose into wharf bank, drops anchor and swings. Then 'drifts' with current alongside wharf.

WHARF: Wellington Dock, Liverpool.   **WHARF AREA** 0.53 ha.   **NGR** SJ 336926

OPERATOR: Mersey Sand Suppliers (Partnership between Norwest Sand and Ballast Co, Tarmac & RMC; Tarmac - formerly Hoveringham, previously ABEL, RMC - previously Coopers). Tarmac operates a concrete batching plant on the wharf, but gravel is brought in by land.

LAND OWNER: Mersey Docks and Habour Co.

SITE DESCRIPTION: Central Liverpool. Small dock within main dock complex. Temporary wharf as far as marine dredged aggregates are concerned.

COMPANY DELIVERING MARINE DREDGED MATERIAL: Norwest Sand & Ballast Co. mv Norstar: vessel chartered to Mersey Sand Suppliers: 850 cargo tonnes.

MATERIALS LANDED: Sand from Mersey estuary and banks off the North Wales coast. Previously sand and gravel landed.

DATE LANDINGS OF MARINE DREDGED MATERIAL COMMENCED: Norwest Sand & Ballast Company's operations started in 1950. The present Partnership started in April 1977. Messrs Abel & Cooper were in business many years before - date unknown.

OWNERS/OPERATORS POLICY TOWARDS WHARF: Wharf is temporary and may be moved at short notice. Minimal current investment in plant, storage facilities etc.

Wharf Case Study No 15

Wellington Dock
Liverpool, Merseyside

Located in the city centre this is a small dock within the main dock complex. It is being used for marine sand on a temporary basis pending a final locational decision by the Mersey Harbour and Dock Company.

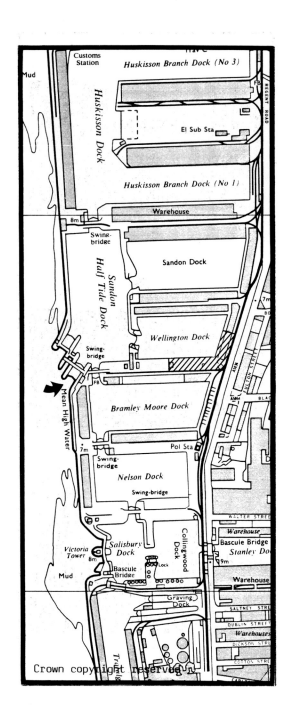

OPERATIONAL CONSIDERATIONS:

Berth: Wharf some 210m long, ship may moor adjacent to any stockpile area.

Docks non-tidal, ship may lock in and out at any time. Ship currently works 2 trips a day, 5 days a week.

Discharge: Takes 3 hours using Poclain grab - swung direct from hold to stockpile.

Wharfage/Plant: No primary processing plant. Material sold as landed. Lorries loaded with front-end loader.

Haulage: provided by sub-contract means.

Stockpile maximum = 8,000 tonnes.

Maximum demand 2,000 tonnes per day, usually 1,000t. Wharf has a limited stocking area.

Material Quality: On-ship screens remove large shells. Chloride levels satisfactory.

Market penetration: Mersey Sand Suppliers deliver around Merseyside, (market area not clearly defined), the rest coming from land pits and beach operations (Southport & Lytham). Haulage based on a loading charge + distance rate. 65 radial km deliveries common; specials (e.g. clay pipe works) up to 100 km. Includes Greater Manchester area.

PLANNING CONSIDERATIONS:

None as in 'dockland' and no adjacent 'sensitive' industries.

COMMENT:

Presently only sand landed, but gravel landed in past and reserves still licensed; gravel landing became uneconomic.

General view of wharf

Plant site

Concrete batching plant

WHARF: Steetley's Wharf.　　**WHARF AREA** 0.3 ha.　　**NGR** NZ 254637
　　　　　Gateshead, Tyne & Wear.　(Excludes new land)

OPERATOR:　　Steetley Construction Materials Ltd.

LAND OWNER:　　Gateshead Borough Council.

SITE DESCRIPTION: Wharf on south bank of River Tyne, in centre of Newcastle/Gateshead. Wharf expanding.

COMPANY DELIVERING MARINE DREDGED MATERIAL:　　East Coast Aggregates Ltd.
(RMC) (normally mv Bowstream, 1800 tonne cargo).

"A Case Study on this wharf was agreed by the Steering Committee (Section 1.1) for inclusion in the report. However, when a draft of this Case Study was circulated to that Committee the Company affected decided to withhold publication of further details."

Wharf Case Study No 16

Steetley's Wharf

Gateshead, Tyne & Wear.

A small wharf located in the centre of Newcastle/Gateshead. Wharf undergoing expansion and modernisation to increase storage and throughput, including conversion to self-discharge from vessels. Strategically located by primary road network.

APPENDIX 4

CODE OF PRACTICE FOR THE EXTRACTION OF MARINE AGGREGATES, December 1981

MESSAGE FROM THE RT HON ALICK BUCHANAN-SMITH MP
MINISTER OF STATE, MINISTRY OF AGRICULTURE, FISHERIES AND FOOD

Both the fishing and the marine section of the sand and gravel industries are engaged in winning valuable resources from the sea. It is in the interests of these two industries, and therefore in the national interest, that the activities of each affect as little as possible the operations and the potential resources of the other. Therefore I warmly welcome the important opportunity this Code of Practice provides for the two industries to work more closely together, in a spirit of co-operation and mutual understanding of each other's problems.

It would indeed be surprising if the difficulties, the misunderstandings and the suspicions that have sometimes existed in the past were to disappear overnight when the Code comes into force on 1 January 1982. But I believe that implementation of the Code should create an atmosphere in which difficulties can be overcome at working level, in which misunderstandings can be avoided and in which mutual suspicions can be allayed by personal contacts.

I welcome the initiative of the Marine Section of the Sand and Gravel Association in promoting such a Code; I thank the Association of Sea Fisheries Committees for agreeing that the Committees they represent should undertake the donkey work of providing local contact points with the fishing industry in England and Wales; and I recognise also the encouragement provided by the Crown Estate Commissioners in the establishment of the Code. I wish the arrangements well.

Alick Buchanan-Smith

FOREWORD

The sand and gravel industry welcomes the opportunity presented by this Code of Practice to improve the understanding of the problems and the system of communication between the fishing and marine aggregate dredging industries.

The winning of sand and gravel from the sea forms an important part of the total national production of such material and features significantly in the structure plans of various regions of the country which are in the process of being approved by the Department of the Environment.

We are aware of the problems and changes being forced on the fishing industry and that proper account must be taken by our members of their legitimate interests and we hope that our willingness to enter into this Code of Practice will be a contributory factor, albeit a small one, to the easing of the pressures on fishermen and their industry.

The Marine Section
Sand and Gravel Association

CODE OF PRACTICE FOR THE EXTRACTION OF MARINE AGGREGATES

INTRODUCTION

1.1 The purpose of this Code of Practice is to provide a basis for close liaison at working level between the fishing and dredging industries in order to promote mutual cooperation and to reduce to a minimum potential interference with each other's activities and damage to each other's resources. It is a voluntary Code which will only be effective if it receives the active support of those directly involved.

1.2 The Code is essentially aimed at establishing practical working communications at local level. Marine extraction is necessary from the national point of view and there will be a need to consider the grant of new dredging licences in areas where there is commercial fishing in order to meet the existing and foreseeable demands for marine aggregates. The object of the Code, therefore, is to assist the development of working relationships between the fishing and dredging industries which minimise interference with fishing while facilitating dredging operations. Modern extraction methods have come a long way from earlier practices and, when properly applied, have a minimum of effect on the seabed.

1.3 It should be emphasised that the establishment of the Code will not affect the formal Government consultation arrangements which are described in the Appendix prepared by the Ministry of Agriculture, Fisheries and Food (MAFF) in consultation with the Department of the Environment (DOE) and the Crown Estate Commissioners (CEC). The Marine Section of the Sand and Gravel Association (SAGA) will continue to have direct access to DOE and CEC on all matters concerning the winning of marine aggregates; liaison with the appropriate fishermen's organisations at national level will continue to be the responsibility of MAFF HQ, which will keep the organisations informed about applications for extraction licences, bulk sampling proposals and prospecting and extraction licences issued for England and Wales.

1.4 The development of liaison between the two industries at working level requires, so far as the fishing industry is concerned, an organised regional network which can offer the dredging companies clear-cut points of contact. In England and Wales, the Association of Sea Fisheries Committees has agreed that this task should be undertaken by Sea Fisheries Committees (SFCs). The Association and the members of the Marine Section of SAGA have undertaken to introduce the procedures set out in the Code on 1 January 1982.

1.5 Most dredging activity takes place off the English and Welsh coasts. However, clear lines of communication are also necessary for operations in Scottish waters. In the absence of SFCs in Scotland, the Department of Agriculture and Fisheries for Scotland (DAFS) will supply individual dredging companies with details of appropriate local contact points for the fishing industry. DAFS will also be responsible for liaising with national fishing organisations. For reasons of clarity the Code contains a separate section for Scotland but the procedures themselves will be essentially the same as those applying to England and Wales.

CODE OF PRACTICE FOR ENGLAND AND WALES

I: PROSPECTING

2.1 The CEC will inform MAFF HQ in confidence when they intend to grant a prospecting licence, with the name of an appropriate company contact.

2.2 MAFF HQ will consult its Fisheries Research Laboratory at Burnham-on-Crouch and the District Inspector of Fisheries, but no outside interests.

2.3 Based on advice from Burnham and the District Inspector, MAFF HQ will notify the CEC (with a copy to the company) of any readily identifiable areas where there might be objections to subsequent extraction. This will be for information only.

2.4 When a prospecting licence is issued, the CEC will inform MAFF HQ, which will send details to Burnham, the District Inspector and SFC. MAFF HQ will provide the company (with a copy to the CEC) with the names and addresses of the District Inspector and SFC (in some cases more than one District Inspector and SFC will be involved).

2.5 Before prospecting starts the company will approach the SFC to discuss working contacts. The SFC will invite the company to provide information on its proposed operations, will supply the company with information about local fishing activity and will discuss any potential difficulties.

2.6 The SFC will inform local fishermen of the issue of a licence and of the company's proposed work schedule and will act as the contact point for any further queries fishermen may have.

2.7 The CEC will inform MAFF HQ of any proposal to take samples by dredging, including the planned area, time and method of working.

2.8 MAFF HQ will inform Burnham, the District Inspector and SFC. If the proposals as they stand would involve a significant risk to fisheries resources MAFF HQ will notify the CEC (with a copy to the company). Such objections will be discussed informally with the company on request.

2.9 If MAFF HQ is satisfied that the proposed operation poses no substantial risk to fisheries resources it will inform the CEC (with a copy to the company), the District Inspector and SFC.

2.10 The SFC will inform local fishermen as appropriate.

II: EXTRACTION

2.11 Under the Government View procedure DOE will consult MAFF HQ on an application for an extraction licence. MAFF HQ will consult Burnham and the District Inspector. This is an inter-Departmental procedure and does not represent a basis for public consultation.

2.12 Unlike prospecting applications, extraction applications are not subject to commercial confidentiality once they have been notified through the Government View procedure. MAFF HQ will therefore also inform the SFC, which will have the opportunity to put views to MAFF HQ.

2.13 MAFF HQ will formulate a Departmental view. If MAFF HQ has in mind to object it will first notify the company informally, through the CEC, giving the reasons. Where possible MAFF HQ and the company will attempt to resolve the objections, involving the SFC where appropriate.

2.14 MAFF HQ will then put its formal comments to DOE under the Government View procedure.

2.15 When an extraction licence is issued, the CEC will inform MAFF HQ, which will send details to Burnham, the District Inspector and SFC. The SFC will inform local fishermen. Wherever possible the working methods listed in Annex A will be used.

III: TIMETABLE

2.16 The timetable for the consultations set out in this Code is at Annex B.

ANNEX A

Code of Practice for England and Wales

WORKING METHODS TO BE USED BY THE DREDGING INDUSTRY WHEREVER POSSIBLE

In response to paragraph 1.2 of the Code of Practice, the dredging and fishing industries will take all reasonable steps to ensure that their activities cause the minimum of interference or damage to each other. In particular, the dredging industry will:-

(a) provide the District Inspector and SFC with as much advance notice as possible of the areas to be worked;

(b) inform the District Inspector and SFC of any areas not being worked for the time being;

(c) work each licensed area in as systematic a manner as is practicable, giving advance information to the District Inspector and SFC of any changes in plan;

(d) wherever possible, work up and down the tidal streams.

ANNEX B

Code of Practice for England and Wales

TIMETABLE FOR CONSULTATIONS

(NB. These are maximum periods and consultations should be completed sooner wherever possible.)

Notification to MAFF HQ of intention to issue prospecting licence

    2 weeks to comment to CEC

Notification to MAFF HQ of issue of prospecting licence

    1 week to notify Burnham, District Inspector and SFC

    1 week to agree working contacts and arrange meeting if needed

Notification to MAFF HQ of proposal to take samples by dredging

    1 week to notify Burnham, District Inspector and SFC

    3 weeks to comment to CEC

Government View procedure

    1 week to notify Burnham, District Inspector and SFC

    4 weeks to comment to MAFF HQ

    2 weeks to formulate MAFF view, arrange meeting with company if necessary, and send final view to DOE

CODE OF PRACTICE FOR SCOTLAND

I: PROSPECTING

3.1 The CEC will inform DAFS HQ in confidence when they intend to grant a prospecting licence, with the name of an appropriate company contact.

3.2 DAFS HQ will consult its Marine Laboratory at Aberdeen, the Sea Fisheries Inspectorate and Inspector of Salmon and Freshwater Fisheries, but no outside interests.

3.3 Based on advice from Aberdeen and the Inspectorate, DAFS HQ will notify the CEC (with a copy to the company) of any readily identifiable areas where there might be objections to subsequent extraction. This will be for information only.

3.4 When a prospecting licence is issued, the CEC will inform DAFS HQ, which will send details to Aberdeen, the Inspectorate and the appropriate fishing organisation or organisations. DAFS HQ will provide the company (with a copy to the CEC) with the names and addresses of the Area Inspector and fishing organisations.

3.5 Before prospecting starts the company will approach the appropriate fishing organisations to discuss working contacts. The organisations will invite the company to provide information on its proposed operations, will supply the company with information about local fishing activity and will discuss any potential difficulties.

3.6 The fishing organisations will inform their local members of the issue of a licence and of the company's proposed work schedule and will act as the contact point for any further queries fishermen may have.

3.7 The CEC will inform DAFS HQ of any proposal to take samples by dredging, including the planned area, time and method of working.

3.8 DAFS HQ will inform Aberdeen, the Inspectorate and appropriate fishing organisations. If the proposals as they stand would involve a significant risk to fisheries resources DAFS HQ will notify the CEC (with a copy to the company). Such objections will be discussed informally with the company on request.

3.9 If DAFS HQ is satisfied that the proposed operation poses no substantial risk to fisheries resources it will inform the CEC (with a copy to the company), the Inspectorate and appropriate fishing organisations.

3.10 The fishing organisations will inform their members as appropriate.

II: EXTRACTION

3.11 Under the Government View procedure DOE will consult DAFS HQ on an application for an extraction licence. DAFS HQ will consult Aberdeen and the Inspectorate. This is an inter-Departmental procedure and does not represent a basis for public consultation.

3.12 Unlike prospecting applications, extraction applications are not subject to commercial confidentiality once they have been notified through the Government View procedure. DAFS HQ will therefore also inform the appropriate fishing organisations, which will have the opportunity to put views to DAFS HQ.

3.13 DAFS HQ will formulate a Departmental view. If DAFS HQ has in mind to object it will first notify the company informally, through the CEC, giving the reasons. Where possible DAFS HQ and the company will attempt to resolve the objections, involving fishing organisations where appropriate.

3.14 DAFS HQ will then put its formal comments to DOE under the Government View procedure.

3.15 When an extraction licence is issued, the CEC will inform DAFS HQ, which will send details to Aberdeen, the Inspectorate and appropriate fishing organisations. The fishing organisations will inform their members. Wherever possible the working methods listed in Annex A will be used.

III: TIMETABLE

3.16 The timetable for the consultations set out in this Code is at Annex B.

ANNEX A

Code of Practice for Scotland

WORKING METHODS TO BE USED BY THE DREDGING INDUSTRY WHEREVER POSSIBLE

In response to paragraph 1.2 of the Code of Practice, the dredging and fishing industries will take all reasonable steps to ensure that their activities cause the minimum of interference or damage to each other. In particular, the dredging industry will:-

(a) provide the Area Inspector and fishing organisations with as much advance notice as possible of the areas to be worked;

(b) inform the Area Inspector and fishing organisations of any areas not being worked for the time being;

(c) work each licensed area in as systematic a manner as is practicable, giving advance information to the Area Inspector and fishing organisations of any changes in plan;

(d) wherever possible, work up and down the tidal streams.

ANNEX B

Code of Practice for Scotland

APPENDIX

TIMETABLE FOR CONSULTATIONS

(NB. These are maximum periods and consultations should be completed sooner wherever possible.)

Notification to DAFS HQ of intention to issue prospecting licence

    2 weeks to comment to CEC

Notification to DAFS HQ of issue of prospecting licence

    1 week to notify Aberdeen, Inspectorate and fishing organisations

    1 week to agree working contacts and arrange meeting if needed

Notification to DAFS HQ of proposal to take samples by dredging

    1 week to notify Aberdeen, Inspectorate and fishing organisations

    3 weeks to comment to CEC

Government View procedure

    1 week to notify Aberdeen, Inspectorate and fishing organisations

    4 weeks to comment to DAFS HQ

    2 weeks to formulate DAFS views, arrange meeting with company if necessary, and send final view to DOE

PROCEDURES FOR LICENSING PROSPECTING AND EXTRACTION OF MARINE AGGREGATES

Introduction

1. Most mineral rights in UK waters are vested in the Crown and companies cannot prospect for, or extract, marine aggregates without a licence from the CEC. In addition, under Section 34 of the Coast Protection Act 1949 operations at sea which may interfere with navigation require the consent of the Secretary of State for Trade. This Appendix describes the formal inter-Departmental consultation arrangements which apply to the issue of prospecting and extraction licences; for convenience it refers only to the arrangements as they relate to MAFF, but DAFS apply very similar procedures in respect of Scottish waters.

Prospecting licences

2. Modern prospecting operations, properly conducted, cause little disturbance to the marine environment or interference with other activities at sea. There is therefore no formal Government consultation procedure and MAFF cannot and does not object to the grant of prospecting licences. The CEC do, however, inform MAFF before issuing any prospecting licence. Having consulted Burnham and the District Inspector, MAFF informs the company concerned of any readily identifiable areas where there may be fisheries objections to subsequent extraction operations. This advice does not prevent the company prospecting in those areas, nor does it automatically imply that there would be no objection to later extraction in other areas. Until they are issued, prospecting licences are regarded as commercially in confidence and MAFF does not consult organisations outside the Ministry. Once a licence has been granted MAFF will provide details to the SFC and the appropriate national fishermen's organisations.

3. Commercial dredgers are commonly used for prospecting operations; however, any bulk sampling must be separately authorised by the CEC, who have agreed to consult MAFF in advance. In the light of advice from Burnham and the District Inspector, MAFF may propose any modifications which it considers necessary to protect important fisheries resources (eg to the time or place of the operation or the method of extraction). Such modifications would be solely concerned with the likely effect of the sampling operation. MAFF would not object to sampling simply on the grounds that there would be objections to full-scale extraction later, although the company would be alerted if there clearly would be such objections. Under the arrangements proposed in the Code there would also be local consultation to ensure that bulk sampling operations did not interfere unnecessarily with fishing activity or fixed gear.

Extraction licences

4. Applications for extraction licences are subject to formal inter-Departmental consultation under what is generally known as the Government View procedure. This is coordinated by DOE, which consults all interested Departments including MAFF. MAFF comments are invariably based on advice from Burnham and the appropriate District Inspector. In addition, the CEC have now agreed that outside interests may be consulted at this stage. If MAFF expects to object to a proposal it will offer the company concerned an opportunity to resolve difficulties through informal discussions before putting comments to DOE. DOE, having considered comments from all Departments, puts a formal Government View to the CEC.

5. The Government View procedure is designed to provide a mechanism for considering licence applications and, if necessary, resolving substantial objections to them wherever possible. Both the fishing and extraction industries are legitimately exploiting the sea's resources; no one industry or activity can have an absolute priority and MAFF does not oppose extraction licences simply on the grounds that the area is fished commercially. Objections are limited to those cases where extraction could seriously damage fisheries resources or interfere to an unacceptable extent with an important commercial fishery.